Waste and Waste Management

T0295395

www.novapublishers.com

Waste and Waste Management

Algal Biorefining: Resource Expenditure and Exergo-Environmental Sustainability
Cynthia Ofori-Boateng, PhD (Editor)
2022. ISBN: 978-1-68507-921-5 (eBook)

What is Biodegradation and Why it Matters
Robert T. Howard (Editor)
2022. ISBN: 978-1-68507-933-8 (Hardcover)
2022. ISBN: 978-1-68507-959-8 (eBook)

Construction and Demolition Waste Management in Australia
Salman Shooshtarian, PhD (Author)
Tayyab Maqsood, PhD (Author)
2021. ISBN: 978-1-68507-237-7 (Hardcover)
2021. ISBN: 978-1-68507-445-6 (eBook)

Pavement Engineering and Waste Materials
Amin Chegenizadeh, PhD (Author)
2021. ISBN: 978-1-68507-238-4 (eBook)

Waste Management: Strategies, Challenges and Future Directions
Nanda Gopal Sahoo, PhD (Editor)
2021. ISBN: 978-1-68507-369-5 (Hardcover)
2021. ISBN: 978-1-68507-394-7 (eBook)

More information about this series can be found at
https://novapublishers.com/product-category/series/waste-and-waste-management/

Adam Fitz
Editor

Municipal Solid Waste Management and Improvement Strategies

www.novapublishers.com

Library of Congress Cataloging-in-Publication Data

ISBN: 979-8-88697-720-2

Published by Nova Science Publishers, Inc. † New York

Contents

Preface

This book is comprised of four chapters focused on municipal solid waste management and improvement strategies.

Chapter 1 - Widespread use of plastics in everyday human life will increase up to approximately 1.25 billion tonnes by 2050. Plastics are generally recalcitrant and remain inert to degradation leading to their accumulation in the environment. One of the biggest challenges of the compost industry is contamination from plastics in municipal solid wastes. Microplastics are found in compost, which can also affect the health of the environment and life forms. Several current methods of screening before and after composting have proven unsatisfactory. The ability of microorganisms to use polyethylene as a carbon source has been recently established. This provides a good opportunity to overcome plastic contamination in compost. However, this can be affected by management, environmental and biological factors. The native microbial communities in different locations can vary with changes in abiotic and biotic factors, which differ from region to region. Therefore, biodegradability of plastics by native microbial consortium needs to be explored per region. Recent research has demonstrated that compost contaminated with microplastics pose a significant danger to soils, wildlife, freshwater, and oceans. These microplastics migrate into other lands and water bodies through surface run-off and wind. It was estimated that when contaminated compost is applied annually between a range of 7 to 35 t/ha in agricultural fields, the plastic load may range between 84,000 and 1.6 million plastic items per ha. This review examines types of plastics and their applications, potential plastic degrading microbes and mechanisms of degradation. It also focuses on recent hypotheses and research findings regarding biodegradation of polyethylene, and identification of potential microorganism(s) which have the capacity to decompose plastics in compost using metagenomics tools. The various challenges posed by plastic degradation and new findings are also reviewed.

Chapter 2 - Inadequate management of waste is gradually turning out to be an issue of concern across the globe due to poor public participation. Participation of members of the public therefore becomes very crucial for a sustainable solid waste management to be achieved. The objective of this paper is to evaluate the perception of the public with regards to the current waste management system in Southwestern Township (Soweto), South Africa as a way of getting the public involved in the implementation of a sustainable solid waste management (SSWM) or Zero waste (ZW) project. To achieve this, a questionnaire consisting of 48 questions was formulated and administered in four communities consisting of informal settlement, middle- and high-income areas. A total of 150 questionnaires were administered but only 118 was returned. Data were collected and analysed using SPSS software with 95% confidence level. Results showed that 51% of respondents were not satisfied with the services, 71% did not know where their collected waste was taken to, for final disposal and 77% of respondents did not know who to contact if they have issues with their waste collection services. From the overall analysis, it was concluded that the people are not properly educated on environmental matters, but they are willing to support Zero waste.

Chapter 3 - This study describes the problems, issues, and challenges of municipal solid waste (MSW) management faced by 26 local authorities in Qalqiliya district of Palestine. Approaches of possible solutions that can be undertaken to improve MSW services are discussed. The study consists of a questionnaire for the public, and a survey with discussions with staff of local authorities involved in waste management. The study provides information on availability of MSW collection services and practices of waste disposal in Qalqiliya district. It was found that little or no consideration of environmental impacts was paid in the selection of MSW dumpsites which were not inspected or monitored consistently. Almost 46% of the local authorities disposed MSW in open random dumps without any treatment, and 15% of them disposed MSW in open random dumps and then burned it. All local authorities offered no pre- or post-employment training to the workers in MSW services, and hence they were usually exposed to serious threats. Small localities shared the workers and vehicles of MSW collection. The number of available waste containers was little in most localities. The average frequency of MSW collection in several localities was 2.2 times per week. It was noticed that <9% of the total budget was allocated for MSW management, making the development of the sector challenging. Results also showed that 97% of the local residents were willing to pay more for better MSW services, 60% of them were willing to separate wastes into organic and inorganic voluntarily, and

19% of them were willing to separate MSW if funded by local authorities. Fortunately, 71.6% of the residents were ready to transform organic wastes into fertilizer products if they were adequately trained. Developing recycling societies does not only alleviate negative environmental impacts, it also promotes the sustainable management of MSW.

Chapter 4 - Food waste generation has continued to be on the increase globally, notably in low-income countries. The estimated global population increases of about 200,000 people daily translates to approximately 100 persons daily for Africa, thus expanding food waste generation. The food waste generated is estimated to be around one-third (approximately ~1.3 billion tonnes) of the total quantity of food produced globally and its economic analysis is valued at ~USD$ 800 billion. Moreover, the amount of food waste generated in sub-Sahara Africa is about 40% of the total tonnes of food produced (~100 million tonnes) and ~10 million tonnes in South Africa yearly. The continuous increase in the amount of food waste generated is now a major source of concern while there are no corresponding facilities to manage these wastes. The current waste disposal method which is landfilling is adequately insufficient and are at the end of its life span, with no intention of replacement due to space scarcity. Thus, this study aims to offer an overview of the issues connected with food waste generation and management in South Africa. Furthermore, this paper aims to summarize the result of acute food waste production on public health and the environment, and the economic and environmental gains of food waste management. Significant analysis of the present methods of food waste management and proposition for an improvement strategy were also discussed.

Chapter 1

Plastics in Municipal Solid Waste Compost: Implications for Farmers, the Environment and Society, and Challenges and Opportunities for Degradation

Seun Esan[1]
Raphael Ofoe[1]
Nivethika Ajeethan[1,2]
Dennge Qin[1]
Sparsha Chada[1]
Bawa Nutsukpo[1]
Matthew Gemmell[1]
Lokanadha Rao Gunupuru[1]
Josephine Ampofo[3]
Samuel Kwaku Asiedu[1]
and Lord Abbey[1,*]

[1]Department of Plant, Food, and Environmental Sciences, Faculty of Agriculture, Dalhousie University, Bible Hill, NS, Canada
[2]Department of Biosystems Technology, Faculty of Technology, University of Jaffna, Kilinochchi, Sri Lanka
[3]Department of Food Science and Technology, University of California, Davies, CA, USA

---

[*] Corresponding Author's Email: loab07@gmail.com.

In: Municipal Solid Waste Management and Improvement Strategies
Editor: Adam Fitz
ISBN: 979-8-88697-720-2
© 2023 Nova Science Publishers, Inc.

Abstract

Widespread use of plastics in everyday human life will increase up to approximately 1.25 billion tonnes by 2050. Plastics are generally recalcitrant and remain inert to degradation leading to their accumulation in the environment. One of the biggest challenges of the compost industry is contamination from plastics in municipal solid wastes. Microplastics are found in compost, which can also affect the health of the environment and life forms. Several current methods of screening before and after composting have proven unsatisfactory. The ability of microorganisms to use polyethylene as a carbon source has been recently established. This provides a good opportunity to overcome plastic contamination in compost. However, this can be affected by management, environmental and biological factors. The native microbial communities in different locations can vary with changes in abiotic and biotic factors, which differ from region to region. Therefore, biodegradability of plastics by native microbial consortium needs to be explored per region. Recent research has demonstrated that compost contaminated with microplastics pose a significant danger to soils, wildlife, freshwater, and oceans. These microplastics migrate into other lands and water bodies through surface run-off and wind. It was estimated that when contaminated compost is applied annually between a range of 7 to 35 t/ha in agricultural fields, the plastic load may range between 84,000 and 1.6 million plastic items per ha. This review examines types of plastics and their applications, potential plastic degrading microbes and mechanisms of degradation. It also focuses on recent hypotheses and research findings regarding biodegradation of polyethylene, and identification of potential microorganism(s) which have the capacity to decompose plastics in compost using metagenomics tools. The various challenges posed by plastic degradation and new findings are also reviewed.

Keywords: compost, plastics, environmental contamination, biodegradation, metagenomics, microorganism

Introduction

The global use of polyethylene and plastic product is approximately 12% per annum, and this continues to rise (Plastic Europe, 2021) The high consumer demand for plastics has driven global production to approximately 140 million tons of synthetic polymers annually, which has increased 1.74-fold within the past 15 years to approximately 243 million tons (Sharma et al., 2015;

Raziyafathima et al., 2016). The rise in production and use of synthetic polymers have increased the amount of global plastic wastes with numerous adverse effects on the environment and society, a concern expressed by the public, including environmental advocates, farmers, compost producers and researchers. Plastic material and its utilization have found wide application in virtually all aspects of human life in both domestic and commercial settings. Thus, hardly will one do without daily encounter with plastics and their products. However, one of the major environmental threats posed by these plastics is their inability to breakdown. This problem can lead to environmental pollution, blockage of waterways and death of marine fresh water flora and fauna (Law, 2017; Blasing and Amelung, 2018). Another negative impact is that plastic contamination in the soil can affect seed germination, plant establishment, root penetration and impede nutrient and water uptake (Hauser and Calafat, 2005; Teuten et al., 2009; Kolodziejek, 2017). Interestingly, there are few published scientific literatures to establish these facts.

Plastic wastes represent at least 16% of the total amount of municipal solid wastes in most landfills (Muenmee et al., 2015). Compost is well-known to be a microcosm of large numbers and diverse populations of microorganisms, which help in the decomposition of organic and some inorganic materials (Koschinsky et al., 2000; Friend and Smith, 2017). Plastic wastes in compost have been labelled as "devil" due to their inert nature and the numerous negative impacts. However, while significant research in biodegradation of plastics has been carried out in soils and plastics, decomposition of plastics in compost is understudied (Sivasankari and Vinotha, 2014).

The diversity and structure of microbial communities can be influenced by abiotic environmental factors and by biotic microbe-microbe interactions (Fierer, 2017). Similarly, compost recipes, preparation methods and composting time play a significant role in shaping compost microbiota (Neher et al., 2013), which can also affect the ability of microorganisms to degrade plastics. Thus, there is a need to investigate the potential of native microbial species in different localities in relation to plastics decomposition in compost. Therefore, this review focusses on the biodegradation of plastic films in compost and the various potential microorganisms previously identified to have the capability of decomposing some of these plastics. This review also discusses the application of next-generation sequencing approaches for evaluation of the factors that shape compost microbiota, and the possibility of using these methods for understanding compost microbiota involved in plastics degradation. The challenges imposed by environment, methodology and strategies are also discussed.

Benefits of Composting and Finished Compost

Composting is an environmentally friendly strategy of organic waste management that has been widely adopted at all levels of governments, organizations, and individuals globally. Compost application is an effective way to add nutrient-rich humus to stimulate plant growth and restore vitality to depleted soils (Earth Easy, 2014). Compost also serves as means of producing organic soil conditioner while the process itself can stabilize, minimize, and eliminate unpleasant odours from decomposing organic matter in the environment, and provide plants with slow-release nutrients that are made available throughout the growing season (USDA, 2010). Compost contains valuable nutrients that could replace or supplement use of synthetic chemical fertilizer while preventing potential health issues that may arise from organic wastes such as Dengue fever, cholera, and malaria (Chandra et al., 2010). The United State Public Health Service has recently identified 22 human diseases that are directly linked to improper municipal solid waste disposal (Alam and Ahmade, 2013). Compost also provides an opportunity to improve the overall waste collection and management programs (Hoornweg et al., 2000). Many researchers have proven that composts harbour communities of microorganisms, (Pérez-Piqueres et al., 2006, Antunes et al., 2016) that are involved in the breakdown of raw organic material into humus-like material leading to the release of energy, carbon dioxide, water vapour and heat (Jeffries, 2003). An effective composting process stabilizes organic carbon (C) and eliminates potential pathogens before the compost is used as a growing medium amendment (Ozores-Hampton et al., 1998).

There are two fundamental types of composting processes - aerobic and anaerobic composting. The aerobic process involves naturally occurring microorganisms that depend on the organic wastes for their energy and body proteins (Fauziah and Agamuthu, 2009). As a result, their activities help to convert the biodegradable organic matter into a humus-like product. This process converts nitrogen (N) from unstable ammonia to stable organic forms. Ultimately, the volume, physical structure and biochemical properties of the waste are altered. The success of the composting process is influenced by many factors such as C/N ratio, temperature, moisture content, oxygen supply, pH and particle size (Guanzon and Holmer, 1993). Anaerobic composting or fermentation is the degradation of organic wastes with very limited oxygen (O_2) to produce methane (CH_4), carbon dioxide (CO_2), ammonia (NH_3), and organic acids (Guanzon and Holmer, 2000). Anaerobic composting is slow compared to aerobic method and involves microorganisms that do not require oxygen to survive, e.g., facultative bacteria.

Plastics

As a modern artificial material, plastics consist of synthetic or semi-synthetic organic polymers and are derived from petrochemicals such as oil and gas via the polymerization of monomers (Wayman and Niemann, 2021). The plastic industry was started in 1907 by Leo Baekeland with the development of polyoxybenzylmethylglycolanhydride (Bakelite), a phenol formaldehyde in New York (Thompson et al., 2009). Later, increased plastic inventions, such as nylon and Plexiglas, occurred during the world war periods. The age of mass production of plastic really began in the 1950s (Chalmin, 2019). The unique intrinsic properties of plastics, characterized by cheapness, durability, and safety, has driven them to become widely used and saturate our world. They have become an integral part of our modern-day life within the last few decades. Currently, around 40% of plastics are designed for single-use applications and packing services, including food, drinks, and tobacco products, to meet the demand for single use properties (Gattringer, 2021).

Statistics on Plastics Production and Use

Over the past 70 years, the mass production of plastics has been accelerated, resulting from the increased perception of need for their vast applications. The global production of plastic was roughly 2 million MT in the 1950s; this has increased significantly to 390 million MT in 2019, an average growth rate of 8.4% annually (Gibb, 2019). The production of plastic is still expanding and is estimated will exceed 500 million MT by 2050 (Sardon and Dove, 2018). Regionally, Asia in 2020 contributed about 52% of the global production of plastics, at 191 million MT. Plastics production in North America and Europe reached about 69 (19%) and 55 (15%) million MT respectively in 2020. It is also worth noting that China has become the leading country in annual plastic production in recent years, accounting for 32% of global production, an estimated 117.5 million MT annually (Plastics Europe, 2021). The consumption of plastic in China amounted to 15.67 kg per capita in 2016. North America consumes more plastic products per capita than any other country globally, estimated at approximately 130.1 kg, which was about 8-fold the quantity of plastics used *per capita* in China.

Environmental Contamination of Plastics

Plastics are ubiquitous and enable our lives to be more convenient and colourful. However, their use is also accompanied by an enormous economic and environmental burden. The recycling rates of plastics are relatively low and limited in scope, reported as only a 9% recycling rate for total plastic production (Gibb, 2019). A large part of plastics ends up in landfills, discarded or eliminated by incineration, resulting in prodigious pollution to air, soil, and water bodies.

Most plastic debris is varied in size and derived from the breakdown of large plastics under physical, chemical, and biological processes (Liu et al., 2021). Presently, microplastic (1 μm-5mm) is considered the most abundant type of debris stored in soil and is present in even greater amounts than found in ocean basins (Liwarska-Bizukojc, 2021). Because of being highly resistant to degradation and increasing usage, microplastics not only may be difficult to degrade, but also could bring significant management problems to the soil ecosystem (Zhao et al., 2021).

Microplastics originate from different sources, including atmosphere deposition, compost, wastewater, and mulch films. Among those sources, mulch films are considered the most critical pathway for entering and contaminating agricultural soils (Alagha et al., 2022, Tian et al., 2022).

The research on microplastic concentration is unevenly distributed throughout the globe. China is one of the countries that has conducted the largest proportion of microplastic research. However, no research has been undertaken to date on microplastic concentration detection in South Asia and Africa (Büks and Kaupenjohann, 2020). Soil microplastic pollution is affected by different factors, such as the land-use type (agricultural base or industrial base), the depth of measured soil samples and the history of land irrigation with wastewater (Zhang et al., 2018, Zhu et al., 2019). These factors make it challenging to compare microplastic pollution on a global scale and make it more suitable to conduct small-scale local studies of microplastic concentrations.

Microplastic concentrations from different sources in different regions are shown in Table 1, where microplastic concentrations are expressed in units of mg of plastic per kg of dry soil.

Table 1. Concentrations of microplastic in soil

Region	Pathway	Land use	Mg/kg	Reference
Nanjing (China)	Mulch film	Agriculture	0.32-0.97	Li et al. (2020)
Shouguang (China)	Mulch film	Agriculture	0.76-1.59	Zhou et al. (2020)
Dianchi (China)	Mulch film	Agriculture	664.5-4020	Zhang and Liu (2018)
Xinjiang (China)	Mulch film	Agriculture	6.55-161.35	Hu et al. (2021)
Mellipilla (Chile)	Wastewater	Agriculture	0.57-12.9	Corradini et al. (2019)
Sydney (Australia)	N/A	Industry	300–67,500	Fuller and Gautam (2016)
Switzerland	Aeolian transport	Floodplain	55.5	Scheurer and Bigalke (2018)
Malmö (Sweden)	N/A	Agriculture	0.3-3.4	Ljung et al. (2018)

Marine plastic pollution is a global issue and has come to prominence in recent years. Plastic waste contributes about 85% of global marine pollution, with 80% of the plastics present in oceans coming from the land and ending up in the ocean environment (Wang et al., 2019). Globally, approximately 14.5 Mt of plastic litter in 2018 was transferred to the ocean. It is anticipated that the total amount of plastic in the oceans will increase three-fold by 2040 and expected to be nearly to 37.7 Mt (Wayman and Niemann, 2021). Plastics can enter the ocean environment via various routes. Ocean plastics from coastline sources originated from municipal solid waste and discarded plastic waste in populated or highly industrialized areas (Halsband and Herzke, 2019). River transportation is also a major pathway; the rivers in Asia accounted for the most significant part of the amount of transferred ocean plastics. In comparison, ocean plastics rarely come from rivers in Africa. Among the Asian rivers, Pasig, the leading plastic polluting river globally, had an input of approximately 62,592 Mt, accounting for 6.4% of the total, followed by Klang and Ulhas, with approximately 13,450 and 13,433 Mt respectively. It is noteworthy that the Philippines has the most plastic polluted rivers globally (Ritchie, 2021). Plastics debris has been widely detected, but the concentration of plastic pollution varies significantly among different marine areas, due to the oceanic surface currents and wind patterns (Tong et al., 2021).

As presented in Table 2, the highest density of ocean plastic debris was found in the North Pacific with 2,000 million pieces. This is followed by 1,300 million pieces in the Indian Ocean and 930 million pieces in the North Atlantic. The Mediterranean Sea has the lowest density with 247 million pieces (Eriksen et al., 2014). Twenty percent of ocean plastics come from marine sources including old fishing nets, traps, pots, and long lines (Watt et al., 2021).

Table 2. Plastic polluted among different marine areas

Marine Areas	plastic pieces (millions)
North Pacific	2,000
Indian Ocean	1,300
North Atlantic	930
South Pacific	491
South Atlantic	297
Mediterranean Sea	247

Plastics can persist in the aquatic environment for extended periods due to their stable physical and chemical properties (Li et al., 2021). These persisted plastics have discernible effects on ocean ecosystems, animal health and even human health due to the high mobility of the ocean environment and the increasing accumulated plastics waste emissions.

Marine animals are helpless when they swallow plastic debris or get tangled in discarded fishing line. Within the past few decades, there have been many news stories about marine animals entangled in plastic lines and discarded fishing gear, causing injury or even death. MacLeod et al. (2021) reported that over 226 species of seabirds, 86 species of marine mammals, all sea turtles, and 430 species of fish had recorded entanglement incidents. In addition, the colorful plastic debris often resembles the food of marine animals, confusing predators, and resulting in more than 100,000 marine animals and 1 million seabirds being killed each year due to plastic ingestion (Walter, 2019).

Plastics in Municipal Solid Waste

Municipal Solid Waste (MSW) is the collective term for the waste, garbage and refuse that we generate every day in urban and rural areas (Nanda and Berruti, 2021). With the population continually rising and the acceleration of industrialization in the world, a rapid increase in MSW accompanies this trend. Approximately 1.9 billion tons of MSW are produced per year globally, with estimates of up to 3.4 billion tons by 2050 (The World Bank, 2020). Globally, the United States is the largest producer of MSW in the world, with approximately 260 million metric tons in 2017. They were followed by China and India, with approximately 220 and 168 million metric tons, respectively (AAAS, 2020). To put this into the perspective of MSW generation per capita, the United States is still the leader in the per capita production of municipal solid waste in the world at 2.58 kilograms per day, whereas China and India produce 1.02 and 0.34 kilograms per day, respectively. Surprisingly, Canada had the second largest municipal solid waste production per capita, about 2.33 kg per day (Visual Capitalist, 2018). Currently, landfills make up the largest share of MSW management globally with 70% in our world, followed by recycling at 19% and energy recovery at 11% (The World Bank, 2020).

The economic and social prosperity of a country or region has been shown to significantly affect the production of MSW. High-income areas tend to have higher living standards, leading to large amounts of MSW (Godiya et al., 2020). MSW comprises different components, including organic waste, miscellaneous, wastepaper, mixed plastics, glass, and metal, as summarized in Table 3 (Nanda and Berruti, 2021). However, these components vary significantly from region to region. In general, organic waste is attributed mainly to the low-income group, while the high-income group demonstrates a high proportion of wastepaper and metal. Both the middle-income and high-income groups have a high proportion of plastic waste (Ramachandra et al., 2018).

Table 3. Global distribution of various components of municipal solid waste

Category	Proportion
Organic waste	46%
Miscellaneous	18%
Wastepaper	17%
Mixed plastics	10%
Glass	5%
Metal	4%

Global recycling rates of municipal solid waste vary significantly among different municipalities worldwide. South Korea and Slovenia had the largest rate of municipal solid waste recycling, at 57% in 2020. This was followed by Germany, at 48%; this was slightly ahead of Australia, at 45% (OECD, 2022). Landfills are still the most common method of disposing of municipal solid waste worldwide because of their lower cost, the cost per ton for landfills depending on their size and increasing as the size of the landfill increases

(Spigolon et al. 2018). In the United States, the average cost of landfill MSW was $53.72 per ton in 2020 (EREF and Waste360, 2021).

Types and Classification of Plastics

Plastics can be defined as polymers that can be moulded into any form or shape when heated (Thakur, 2012). The word plastic comes from the Greek word "plastikos," which means "able to be molded into different shapes" (Crabbe, 1995). Plastics are the most versatile synthetic man-made substances created from fossil fuel resources. They may contain resin materials, which make them very stable and not readily degraded under ambient conditions (Raaman et al., 2012).

Classification of Plastics According to Their Thermal Properties

Common plastics used in recent times are made from inorganic and organic raw materials such as carbon C, sulphur (S), hydrogen (H), nitrogen (N), oxygen (O) and chlorine (Cl)., Plastics are structurally large molecules that are composed of repeated units called monomers with carbon as the backbone (Thompson et al., 2009). Plastics are divided into two distinct groups, thermoplastics (moldable) and thermosetting (not moldable) (Thompson et al., 2009; Ghosh et al., 2013; Henry, 2014). Furthermore, plastics can be classified according to their chemical structure, thermal, degradable, and recycling characteristics. Thermoplastic polymers are those polymers with linear long chain unlinked polymer molecules (Mayer, 2018). They cannot be molded into forms or shapes without the application of heat. However, when frozen they become glass-like and can be easily shattered (Mayer, 2018). They have high molecular weights ranging from 20,000 to 500,000 atom mass unit (AMU) (Singh and Sharma, 2008; Ghosh et al., 2013). Examples are polyethylene, polypropylene (PP), polystyrene (PS), polyvinyl chloride (PVC) and polytetrafluoroethylene. Thermoset plastics such as phenol-formaldehyde and polyurethanes (PU) are synthetic materials that undergo irreversible chemical change when heated but are not recyclable (Singh and Sharma, 2008).

Classification of Plastics According to Their Degradability Properties

Plastics can be categorized into degradable and non-degradable polymers based on their chemical properties. Non-biodegradable plastics are the commonly known synthetic plastics made from petrochemicals (Ghosh et al., 2013). Structurally, they have repeated small monomer units and high molecular weight such as the high-density polyethylene (HDPE). Biodegradable plastics on the other hand are made from starch and are consequently, not very high in molecular weight. They break down easily when exposed to biotic and abiotic factors, such as oxygen, water, microorganisms, sunlight, biodegrading enzymes and high pH (Ghosh et al., 2013; Olaosebikan et al., 2014). A typical example is polybutylene (PBSA). The presence of glycoside linkages and ester groups in the chemical structure of biodegradable plastics represent a point of attack by decomposing enzymes or microorganisms (Reusch, 2013).

Classification of Plastics According to Recycling

In 1998, the Society of Plastics Industry (SPI) established numerical codes ranging from 1 to 7 to allow manufacturers, consumers and recyclers to differentiate between the types of plastics (Earth Talk, 2017). The SPI classifications are described below.

- Plastic (SPI 1): This is the easiest and most common plastics to recycle. They are polyethylene terephthalate (PET) and are accepted in most of the recycling centers.
- Plastic (SPI 2): This group is comprised of high-density polyethylene (HDPE) plastics which can be -used to make laundry detergents, bleach, milk and shampoo containers.
- Plastic (SPI 3): This comprises items made from vinyl. For example, polyvinyl chloride (PVC), which is commonly used in plastic pipes, shower curtains, medical tubing and vinyl dashboards.
- Plastic (SPI 4): This group is comprised of low-density polyethylene (LDPE) plastics used to make thin and flexible plastics like wrapping films, grocery bags, sandwich bags and a variety of soft packaging materials. It is common in all the landfills.

- Plastic (SPI 5): This group consists of polypropylene (PP). Some food containers and most plastic cups are made from this plastic.
- Plastic (SPI 6): This group is comprised of Polystyrene (PS) plastics commonly called styrofoam. Examples include items such as coffee cups, disposable cutlery and meat trays. PS can be reprocessed into many items including rigid insulations.
- Plastic (SPI 7): Plastics under this category are difficult to recycle. They comprise other types of plastics or combinations of two or more types. Examples are compact discs and medical storage containers.

Plastics in Compost: Why Is It a Problem to Composters?

As shown in Figure 1, plastics in compost have extensive implication to composters and their business, environment, agroecological systems, plants and society. During the recycling of organic wastes through composting, plastics in the waste materials do not fully degrade and become nuisance contaminants in the final compost product (Mistry et al., 2018; Weithmann et al., 2018). Germany has strict regulations on the quality of organic fertilizer, which allows up to 0.1% by weight of plastics in compost. In this regulation,

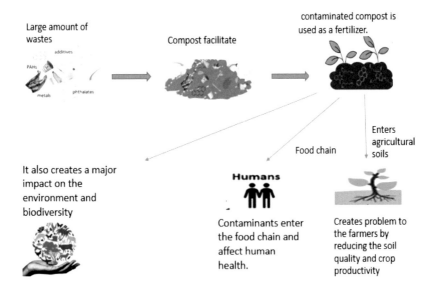

Figure 1. Problem caused by plastics in compost to composters, environment, animals, farmers, public health, and biodiversity.

plastic particles smaller than 2 mm are not acceptable (Weithmann et al., 2018). Hence, compost manufacturers, spend time sorting out these plastics before composting or removing the particles left in the compost after processing. These activities increase the operating costs and the contamination degrades the quality of compost produced, making the composters less economically viable and undermining the broader sustainability of the composting industries (Mistry et al., 2018; Association of Recyclers of Oregon, 2020). Folino et al., (2020) reported that compost manufacturers spend a significant amount of time processing waste materials containing plastics. The production of low-quality compost due to the presence of plastic particles hinders the sale of the compost, making it either difficult to sell or resulting in the consumers not buying it at all (Association of Recyclers of Oregon, 2020).

Problem to Environment

Brinton et al., (2018) suggested composting as a key method for reducing methane accumulation in landfills and leading many communities toward zero waste. However, for this degree of composting to develop in the United States and Canada, much attention must be given to the quality of the composts disseminated into the overall environment, since this will greatly impact local and global ecosystems. Eco-Cycle and Wood End Laboratories conducted research in 2010 to ascertain how plastics react when used as feedstock in composting and the result has shown that these plastics are shed into fragments and contaminate the finished product. Another study, conducted in 2011 and reported by Platt et al., (2014), also showed that plastics used in composting produce macro and micro fragments that contaminate finished compost. Based on this research, Cedar Grove and King County municipal governments in Ontario banned plastic-coated paper products from being used in their composting systems (Brinton et al., 2018).

Plastics are ubiquitous pollutants, found in the atmosphere, marine, freshwater and terrestrial environments and inside living organisms (Scopetani et al., 2022). Plastic fragments are released into the environment when soil amendments containing plastic fragments are applied to agricultural lands (Weithmann et al., 2018; Wan et al., 2019). The macro and microplastic particles that contaminate the compost do not degrade and are dispersed into the environment causing detrimental effects on the ecosystems (Platt et al., 2014; Leal Filho et al., 2019). These plastic particles migrate into other lands and water bodies through surface run-off and wind (Brinton et al., 2018; Leal Filho et al., 2019). These plastic particles could also be found on beaches in many countries (Gajšt et al., 2016; Brinton et al., 2018).

Microplastic particles have been demonstrated to dissolve into nano plastic particles, generally defined as less than 100 or 1,000 nm in at least one of its dimensions, in the environment (Brachner et al., 2020). When compost is applied between a range of 7 to 35 tons per hectare in agricultural fields, and from 6.48 to 19.44 tons per hectare in horticultural soils, plastics load may range between 84,000–1,610,000 and 77,770–894,240 plastic items per hectare yearly, respectively (Scopetani et al., 2022). Research conducted by Scopetani et al. (2022) on plastic fragments in compost and the transfer of dangerous contaminants from plastics to compost and soils confirmed that compost usage is a source of plastic pollution in agricultural fields and poses danger to the soils. Weithmann et al., (2018) reported the presence of up 895 particles of plastics in a kilogram of compost. The Association of Recyclers of Oregon, (2020) revealed that the plastics used during composting contains toxic chemicals that contaminate the finished compost and when this contaminated compost is applied to the soils, it ends up polluting the environment. Plastic materials contain pigments, plasticizers, flame-retardants, antioxidants and other additives that cause damage to the ecosystems and human health (Hahladakis et al., 2018). Gajšt et al., (2016) and Scopetani et al., (2022) also reported that plastic fragments serve as a reservoir for toxic chemicals in the environment.

Problem to Animals

Microplastics appear to be easily uptaken and digested by soil animals as they enter soil ecosystems (Zhang et al., 2020). Plastic fragments larger than 5 mm dispersed in the ecosystems can be ingested by a wide range of animals, including fish, birds, turtles and cetaceans. When predators consume prey, these plastic particles are transferred into their systems through the food web (Gajšt et al., 2016; Brinton et al., 2018; Lael Filho et al., 2019). The ingestion of microplastics by aquatic animals could pose detrimental effects to their health and survival (Zhu et al., 2019). Plastic particles can negatively influence the feeding, reproduction, survival, growth, and immune systems of soil organisms (Zhu et al., 2018). For instance; microplastics negatively affect the survival of nematodes in the soil (Zhu et al., 2019), reduce the growth of earthworms (Huerta et al., 2016; Qi et al., 2018), cause high mortality in earthworms (Huerta et al., 2016), and inhibit the growth and reproduction of *Folsomia candida* in the soil. The downward movement of plastics in the soil profile may cause toxicity to earthworms (Zhang et al., 2020). Plastic fragments in compost, when applied to the soil, could block the pores in the soil thereby restricting the movement of soil animals and even changing their ovipositional sites, which could indirectly influence their reproduction. Plastic particles could reduce the food intake of snails, springtails, and nematodes (Zhang et al., 2022). They could also block the digestive tracts of the soil microorganism, hence, influencing the gut microbiota and changing their digestive function (Fackelmann and Sommer, 2019). Microplastics could enter the circulatory system of marine fish through their intestinal walls; marine fishes have been found with microplastics in their muscle tissues (Zhang et al., 2022).

Problem to Farmers

A greater knowledge gap exists when it comes to the quantity of plastics and associated contaminants that agricultural fields receive from the use of recycled fertilizers, such as bio waste and sewage sludge compost (Bläsing and Amelung, 2018; Weithmann et al., 2018; Zhang et al., 2020; Braun et al., 2021). However, microplastics in soil are unlikely to be taken up by plant roots. Nano plastics created by the breakdown of these microplastics may be taken up by plants (Golwala et al., 2021). Plastic fragments in soil amendments can have a direct impact on the health of the soil to which they were applied, by altering the characteristics of the soil, such as texture and structure. This influences plant performance after contaminated composts have been applied to the farm lands.

Plastic particles in compost affect the effective production of farmers' crops. The presence of plastic fragments in the soil impedes the growth of wheat (Qi et al., 2018), maize (Boots et al., 2019; Hu et al., 2020), rice (Dong et al., 2020), lettuce (Gao et al., 2019). Biomass of fruit was also negatively affected by these microplastics in compost (Zhu et al., 2019). It is believed that plastic particles are too large to pass through the physical barriers of intact plant tissue. In contrast, a few facts disprove this assumption. Microplastics were found in carrots, lettuce, broccoli, potatoes, apples, and pears in Italian retail vegetables. The researchers discovered the most microplastic contamination in apples and the least contamination in lettuce and hypothesized that the perennial nature of a fruit tree permitted microplastics to collect more than in annual crops. Overall, vascular plant studies show that microplastics can be absorbed by roots and delivered to epigeous organs and tissues via bottom-up transport, particularly for smaller nano-sized microparticles (Agathokleous et al., 2021). Microplastics can also be absorbed by roots and transferred to vascular plants' shoots and leaves via the xylem pathway (Li et al., 2020). The amount of absorption entirely depends upon the plant anatomy, composition, and shape, with the following absorption rates noted: polyvinyl chloride microplastics (100 nm-18μm) in lettuce (*Lactuca sativa*) (Li et al., 2020) and polystyrene microplastics (88 nm) in wheat (*Triticum aestivum*) plants (Lian et al., 2020).

Problem to Public Health

Plastic fragments in compost are not only destroying animals and plants, but also endangering the health of the human population by threatening food security (Leal Filho et al., 2019; Igalavithana et al., 2022). Microplastics have become ubiquitous in human food stuff and their intake poses a potential threat to human health (De-la-Torre, 2020). Toxic chemicals in the plastics released in the compost are transferred into the soils and water bodies. These toxic chemicals are taken up by plants and lead to negative health impacts when products from these plants are consumed by the public (Association of Recyclers of Oregon, 2020). Microplastics enter the food web due to their microscopic size and have been identified in human food, particularly seafood (Toussaint et al., 2019). Bradney et al., (2019) revealed that the consumption of contaminated seafood causes health problems to humans. Organic (chromophores) and inorganic (metals) pigments, which are used to make various coloured plastics, can harm the humans neurological and reproductive systems (Campanale et al., 2020). Microplastics with a diameter of less than 130 micrometres can accumulate in human tissue as a result of cellular and paracellular transport, among other mechanisms. They can produce poisons, chemicals, and monomers that have been linked to carcinogenic tendencies once they invade. Furthermore, human exposure to microplastics can result in lung damage and alterations in liver function (Karimifard and Moghaddam, 2018). Some other human health concerns include upper respiratory issues and autoimmunity diseases (Bradney et al., 2019).

Problem to Biodiversity

Biodiversity is the vast variation of all forms of life on earth, such as, the kind of plants, animals, microorganisms, their genetic make ups and the ecosystem they comprise (Rawat and Agarwal, 2015). Biodiversity is grouped into three major levels: genetic diversity, species diversity and ecosystem diversity (Rawat and Agarwal, 2015). Biodiversity is the system behind the performance of an ecosystem, especially in the communities of above ground organisms (Wagg et al., 2014). All living organisms depend on biodiversity for material well-being, including clean air, food, fresh water, natural resources and medicine (Rawat and Agarwal, 2015; Population Matters, 2021).

Despite the numerous benefits of biodiversity, it is threatened by agriculture intensification, pollution, soil compaction and sealing, soil acidification, deforestation, urbanisation, loss of organic matter, salinization and sodification, fire, erosion, landslides, climate change and invasive species (Population Matters, 2021, Köninger et al., 2022). The loss of biodiversity and the resulting environmental changes are currently manifesting quicker than ever in human history, and there is no sign that this trend will slow down (Rawat and Agarwal, 2015). Microplastic pollution is expected to rise in future decades, which poses a significant threat to biodiversity (Agathokleous et al., 2021). Zhang et al., (2022) reported that microplastics are becoming the main pollutants which have been widely detected in agricultural ecosystem. Composting introduces plastic pollution into agricultural areas, with the plastic potentially transferring dangerous pollutants to the biodiversity and soil ecosystems (Scopetani et al., 2022).

Microplastics can be ingested by soil animals and excreted rapidly, sometimes in less than an hour, through feces (Agathokleous et al., 2021; Zhang et al., 2022). Agathokleous et al., (2021) revealed that the excretion of ingested microplastics is significantly influenced by the particle size and shape and exposure or depuration duration. Smaller particle size may persist longer in the cells, while larger fragments tend to be egested more quickly (Okubo et al., 2018). Agathokleous et al., (2021) and Zhang et al., (2022) reported that the ingestion of microplastics causes toxicity, oxidative stress, weak nutrition supply, intestinal damages and other adverse effects. Earthworm galleries can also transfer microplastic-associated contaminants into the rhizosphere and deeper soil layers, changing the structure of vascular plant communities and soil microbial communities. Microplastic pollution may alter communities and ecosystems, providing complicated risks to biodiversity, according to similar results of possibly impacted composition, structure, and function of numerous microbial communities noted in various aquatic and terrestrial systems (Agathokleous et al., 2021).

Degradation of Plastics in Compost

Degradation of plastics can be viewed as changes in the physical and/or chemical properties that are caused by environmental factors including moisture, heat, light, and biological or chemical activities. The process of plastic degradation has been characterized by several authors and categorized

Table 4. Common mode of plastic degradation

Factors	Photo-degradation	Thermo-oxidative degradation	Hydrolytic degradation	Biodegradation
Active agent	UV-light or high-energy radiation	Heat and oxygen	Water and oxygen	Microbial agents
Requirement of heat	Not required	Higher than ambient temperature required	Not required	Not required
Rate of degradation	Initiation is slow. But propagation is fast	Fast	Very slow	Moderate
Other consideration	Environment friendly if high-energy radiation is not used	Environmentally not acceptable	Environmentally not acceptable	Environment friendly
Overall acceptance	Acceptable but costly	Not acceptable	Acceptable but costly	Cheap and very much acceptable

Modified from (http://www.envis-icpe.com, Plastics recycling-Economic and Ecological Options. ICPE 2006;4(4):1–12).

into chemical, physical and biological reactions that result in the degeneration of function or the formation of new structurally homogeneous functional groups (Pospisil and Nespurek, 1997; Tokiwa et al., 2009). Generally, plastic can be degraded via mechanical, hydrolytic degradation, photodegradation, thermo-oxidative degradation and biodegradation mechanisms (Gewert et al., 2015; Andrady, 2011). Natural degradation of plastic begins with photodegradation which results in thermo-oxidative degradation. This generates simple carbon molecules that can be used -by microbes as energy sources for further degradation (Andrady, 2011). The widely reported mode of action of plastic degradation is summarised in Table 4 above.

Photodegradation

Photodegradation is viewed as the most effective plastic degradation pathway (Gewert et al., 2015). Photodegradation of plastics involves the ability of plastics to absorb harmful radiation from the sun. The ultraviolet light from the sun which include UV-A (~315–400 nm) and UV-B radiation (~295–315nm) have been reported as the primary factors facilitating photo-degradation on plastics in terrestrial soils (Raquez et al., 2011; Andrady, 2011; Liu et al., 2020). To photodegrade a material, the material must contain or interact with chromophores that absorb photochemically active sunlight (Stubbins et al., 2021). Moreover, the visible part of sunlight radiation promotes degradation by heating, while infrared radiation facilitates thermal oxidation of plastics (Webb et al., 2012). Ultraviolet radiation from the sun provides the minimum energy required to instigate oxygen incorporation into plastics. Most plastics absorb this high-energy radiation at a specific spectrum, which activates their electrons to higher reactivity. This causes oxidation and makes plastics become brittle and break into smaller pieces, until the polymer chains reach a sufficiently low molecular weight to be further metabolized by other soil agents, including microorganisms (Liu et al., 2020; Andrady, 2011). In a more recent study, Ding et al. (2022) revealed that light radiation treatments (wavelength range 295-2500 nm), with and without the presence of soil processes, facilitated noticeable alterations of PET microplastics, including wrinkles and cracks. Additionally, such alterations were more prominent on the PET surfaces, as the time of photodegradation increased and they were faster and more intense in the presence of soil processes than without soil processes.

Intriguingly, the degree of photodegradation is negatively regulated by increasing organic carbon content in soils (Ding et al., 2022; Chen et al., 2019). While compost is known to contain a high amount of organic carbon (> 50%), the rate of photodegradation of plastics can be minimized or inhibited. The possible reason for this phenomenon could be that organic carbon can form noncovalent bond planar with organic compounds via chelation and hydrogen bond interactions. This could increase the half-lives of both due to UV shielding and the stabilizing effect of chemical binding and may slow down the photolysis of plastics (Ding et al., 2022; Liu et al., 2020; Chen et al., 2019; Lagalante et al., 2011; Qu et al., 2017). Although organic carbon content could decrease the photodegradation of plastic, this may vary depending on its amount. For example, Ding et al. (2022) found a high photodegradation rate in five different soils, possibly because the inhibitory effect becomes not significant when the organic matter content reaches a certain value. A similar effect was observed with the photodegradation of BDE plastics by Liu et al. (2020). Thus, it can be concluded that the association between compost properties and photodegradation rate suggests that compost components have a great influence on plastic removal.

Thermal Degradation

Thermal degradation of plastics is the most common mechanism used worldwide. It is sometimes referred as incineration of plastics and allows for recovery of energy form plastics. While this mode has stirred up public concerns in most countries due to pollution of the environment, it is effective in reducing plastic waste masses (Rujnić-Sokele and Pilipović, 2017; Ali et al., 2021). Thermal degradation of plastics is a molecular deterioration that results from overheating of plastics at a high temperature. At such temperatures, the integral components of plastics, which is characterised by a long chain backbone of polymers, begin to split, and react with one another to alter the features of plastics. These changes involve alteration in molecular weight, and physical and optical property changes, which includes colour change, reduced ductility, and embrittlement, chalking and cracking (Shah et al., 2008). However, in most composting facilities, thermal degradation of plastics is noticed during the initial composting process, which is conside3red as an environmentally friendly means of waste management. Several studies revealed that thermal degradation of plastics varies between polymers and incubation temperatures. For example, Al Hosni et al. (2019) revealed that polycaprolactone (PCL) exhibited a faster degradation rate when buried in compost and incubated at 50°C after 91 days. Similarly, PLA can be fully degraded when the compost temperature reaches 60°C (Shah et al., 2008), while degradation of PBS was noticed to be higher in compost compared to soils due to increased temperature and humidity (Kim et al., 2005).

Biodegradation

Some microorganisms such as fungi, bacteria, actinomycetes and few algae involve the degradation of natural and synthetic plastic materials through their metabolic activity, this process is called as "biodegradation." Biodegradation pathway and dominant microbial groups are often determined by the environmental factors, such as the presence of oxygen (Shah et al., 2008). Where oxygen is present, for example, in the soil composite method, aerobic microorganisms are mostly responsible for degradation of complex materials, and they form carbon dioxide (CO_2) and water (H_2O) with microbial biomass as their end products. This process is called aerobic biodegradation (Venkatesh et al., 2021). In contrast, under anaerobic conditions, anaerobic consortia of microorganisms are responsible for polymer degradation. Mostly in landfills and sediments anaerobic biodegradation leads to methane (under methanogens) (Barlaz et al., 1989), H_2O and CO_2 with microbial biomass as their end products. Hydrogen sulfide (H_2S) is produced under the existence of sulfate-reducing bacteria.

Usually, it is a complex process to produce these types of end products from the long molecular weight polymer, which needs various microbial activities in each step. At each level, a particular microbial community breaks the polymer into granules and another one utilizes the monomers and excretes them (Shah et al., 2008). Biodegradation is influenced by various factors, such as the type of dominant microorganism, characteristics of the polymer and the nature of the pre-treatment. The polymer characteristics play an important role in its degradation that includes polymer mobility, tacticity, crystallinity, molecular weight, the type of functional groups and substituents present in its structure, and plasticizers or additives added to the polymer (Artham and Doble, 2008). Generally, larger molecular weight results in a lower rate of polymer degradability by microorganisms. In contrast, monomers, dimers, and oligomers of a polymer's repeating units are much more easily degraded and mineralized.

Factors Affecting Biodegradability of Plastics

Generally, biodegradation of plastics is linked to their properties and the physical and chemical composition critically influence the biodegradation mechanism. The factors can be broadly grouped into three categories: environmental, polymer characteristics and microorganisms. Moreover, the chemical composition of plastics acts as a primary factor that influences its ability to degrade. This includes surface conditions (hydrophobic and hydrophilic properties, surface area), intrinsic structures (molecular weight, functional groups, molecular weight distribution) and morphological features (crystallinity, degree of elasticity, tacticity), which greatly influence the melting and glass transition temperature of plastics (Luyt and Malik, 2019; Tokiwa et al., 2009). Additionally, environmental conditions, including temperature, oxygen, light and pH are important factors and, moisture availability must be proportional to the properties of the plastics to initiate and facilitate the biodegradation process.

Polymer Characteristics

Several studies have revealed that the level of molecular weight determines the ease of degradation, as it predetermines many physical properties of plastics. Molecular weights plastics composed of simple units, including monomers, dimers, and oligomers, are easier to degrade in compost than polymers with high molecular weights with complex networks of polymer chains that require high depolymerising energy (Kale et al., 2007; Gu et al., 2017). For instance, polyhydroxybutyrate (PHB) is a naturally occurring plastic, which is composed of small monomer units and can be synthesized from cellulose and starch. However, in compost, PHB can be degraded in five to six weeks by a range of bacterial species through the action of a PHB depolymerase enzyme (Tokiwa et al., 2009). Some studies suggested that LDPE plastics in compost could undergo varying levels of thermal and photo-degradation (Andrady, 2011), which could result in the production of low molecular weight compounds. Also, PCL with a higher molecular weight (> 4000 Da) was slowly by degraded by a lipase from *Rhizopus delemar*. while degradation at a molecular weight lower than 4000 Da was more significant (Tokiwa and Suzuki, 1978). Similarly, the surface erosion of non-oxidized LDPE with a low molecular weight (1700-23700 Da) was more prominent, which further suggests that the biodegradability of plastic increases with decreasing molecular weight. The degree of crystallinity is a significant factor that influences the biodegradability of plastics. Plastics with an amorphous structure are more susceptible to degradability than crystalline ones (Tokiwa et al., 2009). The ease of biodegradability of amorphous plastics is facilitated by their loosely packed chains that allow water permeability. However, crystalline plastics are tightly packed and impermeable to water, rendering them more resistant to biodegradability.

Studies have revealed that the rate of plastic degradation reduces with increasing crystallinity (Jenkins and Harrison, 2008). For example, LPDE plastics are less crystalline, due to their side branches and it has been confirmed experimentally that the first amorphous regions are consumed (Santo et al., 2013; Sen and Raut, 2015). Similarly, Chamas et al. (2020) compared the half-lives of different plastic polymers using pseudo-zeroth-order kinetics and showed that the specific rate of PE degradation (relatively more crystalline) was 9.5 μm year-1 compared to 1105 μm year-1 of PET under the same conditions. The flexibility and rotatability of plastics have been reported to influence degradation (Kijchavengkul and Auras, 2008). PGA plastics could degrade faster than PLA due to the presence of methyl side chains groups in PLA, thus rendering PLA plastics less flexible (Kijchavengkul and Auras, 2008). Also, aliphatic polyesters are easier to degrade than aliphatic-aromatic and aromatic plastics. Aliphatic polyesters with higher polymer flexibility have higher degradability while aromatic polyesters with lower polymer flexibility are not biodegradable, implying that the rigidity of aromatic polyesters requires higher energy to degrade. The hydrophobicity of plastics has been shown to significantly affect its degradation rate (Lim and Thian, 2021; Chamas et al., 2020). An increase in hydrophilicity of polymers increased their susceptibility to degrade via the action of enzymes. Hydrophilic surfaces possess a high surface energy, which reduces the contact angles with water and promotes microbial attachment and facilitates the degradation process.

The availability of functional groups can increase the hydrophilicity of plastics. Thermal-oxidation and/or UV pre-treatment of polypropylenes can result in the formation of carboxyl, carbonyl and ester groups and increase hydrophilicity of polypropylenes. Fusion of hydrophobins into cutinase from *T. cellulosilytica* enhanced the hydrophilicity of PET plastics by 16-fold (Ribitsch et al., 2015). Min et al. (2020) demonstrated, using molecular dynamic simulation, that polypropylene plastics with high hydrophobicity exhibit the least biodegradative potential, compared to a nylon with a relatively high hydrophilicity. Furthermore, glass transition (T_g) and melting temperatures (T_m) are secondary plastic characteristics that strongly affect the rate of microbial degradation. It has been previously revealed that an inverse relationship exists between T_m and biodegradation rates. Nevertheless, plastics T_m is affected by changes of melting enthalphy (ΔH) and entropy (ΔS), as shown in the equation; $T_m = \Delta H/\Delta S$ (Tokiwa et al., 2009). Plastics with high Tg are glassier and less flexible (Kijchavengkul and Auras, 2008), whereas plastics with low Tm exhibit a higher rate of degradation when hydrolyzed by *R. delemar* lipase. These properties indicate that the presence of such materials in compost feedstock will affect the quality and overall use in crop production.

Environmental Factors

The effect of individual and/or interaction of several environmental elements on plastics are crucial in the kinetic dynamics of biodegradation by microbes. Such environmental elements, including light, moisture, temperature, oxygen, light and pH and biological activities have been found to promote hydrolysis of polymers (Siracusa et al., 2019). The availability of moisture enhances the miniaturisation of plastics via an increase in solubility and hydrolysis rate (Lim and Thian, 2021; Chamas et al., 2020).

Moisture is crucial for growth and proliferation of microbes and essential elements in the hydrolytic process of polymer chain reduction. As a result, microbial activities are supported, which facilitates more polymer chain scission, thereby, promoting biodegradation. Several plastics have been noticed to be sensitive to UV radiations (Lim and Thian, 2021; Chamas et al., 2020). Ultraviolet radiations emitted from sunlight are readily available as an energy source that degrade plastics. Such radiations carry photons that elicit excitation in polymer chains and result in the production of radicals that trigger the oxidation process and lead to polymer chain scission (Singh and Sharma, 2008; Brebu, 2020). Thermal bio-degradation has also been observed to occur solely, and sometimes simultaneously, with UV radiation. Photodegradation does not require additional energy above the environmental temperature to facilitate degradation. Additionally, the photodegradation process decreases exponentially with decreasing light intensity and increasing wavelength in several common plastics and polymers (Singh and Sharma, 2008; Lim and Thian, 2021).

High temperatures have been observed to cause thermal degradation of plastic and such temperature can only be raised up to a point, as some microbe activities decrease at too high a temperature and enzymatic degradation activities cease. Temperatures in composting materials have been estimated at 55°-60°C, a condition which promotes degradation rates (Hosni et al., 2019). In landfills, such temperatures have been estimated at 100°C, which provides sufficient moisture and oxygen for subsequent degradations via thermal-oxidative and hydrolytic mechanisms (Hao et al., 2017). High temperatures increase the kinetic energies of atoms, which causes distorting in molecular structure leading to molecular scission of long chain backbones of plastics (Ray and Cooney, 2018). Oxygen is a non-discretionary component in aerobic degradation (Ali et al., 2021). The presence of high levels of oxygen accelerates the degradation of plastics by facilitating chemical reactions between oxygen-carbon centered radicals produced during the initial stages of the degradation processes (Lim and Thian, 2021).

Additives and Reagents

Recently, additives, including stabilizers or plasticizers added during the production of plastics which act as pro-oxidants, flame retardants or pro-degradants to strengthen plastics properties, have been shown to greatly affect their degradation rate. Such additives function by activating, inhibiting, or catalysing the biodegradation process by influencing the wettability of the functional groups and causing a significant reduction in plastics degradation (Aldas et al., 2018). Some additives have been shown to inhibit microbial degradation. For example, dibutyl tin dilaurate is a highly toxic plastic additive in PU with antimicrobial effects and reduces microbial activities on plastics (Cregut et al., 2013). Similarly, the use of simple and alternative carbon sources has been shown to regulate microbial activities. This effect was observed when the degradation of PE derivatives by *Pseudomonas* strain was increased by 80% upon the removal of glucose in media (Tribedi et al., 2012). In contrast, the addition of starch and palmitic acid as an additive acted as a nutrient source and enhanced the biodegradation of LDPE (Mehmood et al., 2016) and polyethylene plastics (Jayaprakash and Palempalli, 2018), respectively. Oxidative agents, including sulphuric acid, hydrochloric acid and nitric acid, have been demonstrated to oxidize plastics (Moharir and Kumar, 2019) while surfactants such as Tween 80, increases the hydrophilicity of plastics and promotes microbial degradation (Ghatge et al., 2020).

Plastics Degradation by Microorganisms

The biodegradation of polymer mechanisms is comprised of three steps; (a) as a first step the microorganisms attaches on the surface of the polymer (Attachment), (Gu, 2003) (b) in the second step, the polymer can be utilized by microbes as a carbon source (Utilization) and (c) polymer degradation which incorporates biodeterioration, bio-fragmentation, mineralization, and assimilation. *In compost, microbial* communities are a valuable source of enzymes with *degrading* activities on synthetic polymers. (Wei and Wierckx, 2021). The surface of the plastic acts as the attachment site for microbes and then they degrade these polymers by secreting enzymes to obtain energy for their growth and survival (Danso et al., 2018). Due to this process, large molecular weight (large) polymers are degraded into low molecular weight molecules, such as monomers and oligomers. Some oligomers may be assimilated in the internal environment of microorganisms after diffusing inside them.

Mechanism of Microbial Biodegradation

The complete process of microbial plastics degradation includes bio-deterioration, bio-fragmentation, mineralization, and assimilation.

Biodeterioration

Biodeterioration is sometimes described as a surface-level degradation. It is the process that changes the surface of plastics and modifies their chemical, physical and mechanical properties. For example, the presence of a colony of bacteria, fungi, or algae on the surface of a plastic can give rise to an aesthetic deterioration in the properties of the plastics and it could resulting in the appearance of dark fungal colonies, which would make the plastic look "dirty." All chemical and structural changes depend on the type, structure, and composition of polymers. Biofilm formation and substrate formation inside the plastic are due to the process of deterioration (Vivi et al., 2019). While biodeterioration typically occurs as the first stage of biodegradation, it can, in some cases, be parallel to bio-fragmentation (Muller, 2005). This step involves the secretion of adhesives or biosurfactants.

Bio-Fragmentation

During the bio-fragmentation step, enzymatic actions evolve on the plastic polymers. This involves hydrolysis and/or fragmentation of the polymer carbon chains and the release of intermediate products, mediated by enzymes secreted by microorganisms. Oxygenases, mostly contained enzymes in bacteria, could break oxygen molecules, which are then added to the carbon chains and as a result, alcohol and peroxyl products are formed that are less harmful (Pathak, 2017). The transformation process of carboxylic groups is catalyzed by lipases and esterases or by endopeptidases for amide groups (Elahi et al., 2021).

Mineralization

During this step, plastic polymers, which are formed in the bio-fragmentation process, enter the microbial cells through cell membranes. Monomers that are large in size cannot be transferred to the inside of the cells and tends to stay outside (Elahi et al., 2021). The small monomers that moved inside the cells are oxidized and utilized for the energy production. This energy eventually is used for biomass production (Kale et al., 2015).

Assimilation

Small hydrocarbon fragments released by bio-fragmentation are taken-up and metabolized by bacteria or fungi. During the assimilation process, atoms are integrated in the microbial cells for complete degradation. Secondary metabolites are transported outside the cells or transferred to other microbes that further perform degradation and use these metabolites (Elahi, *et al.*, 2021). The oxidized products, such as CO_2, N_2, H_2O and CH_4, are released during metabolite degradation.

Plastic Degradation by Bacteria and Fungi

There are various studies reported on plastics degradation by microbes (Table 2). Iiyoshi et al. (1998) and Kim et al. (2005) reported that fungi and various *Streptomyces* sp. Bacteria, such as *S. setonii* 75Vi2, *S. viridosporus* T7A and *S. badius* 252, secrete extracellular enzymes that aid in the decomposition of plastics. For example, in fungi, the ligninolytic system's extracellular enzymes comprise laccases, oxidases, enzymes and catalases that produce the extracellular hydrogen peroxide (Ruiz-Duenas and Martinez, 2009), which leads to the plastic degradation. Aswale and Ade (2009) reported the biodegradation of carry bags by bacterial isolates. Two actively involved strains (*Pseudomonas sp.* and *Bacillus cereus*) were identified by using biochemical tests and morphological keys in their study and degradation of plastic polyethylene by lignocellulose degrading fungal *Phanerochaete chrysosporium* was reported. Aswale and Ade (2009) evidenced that *Serretia marcescens* degradation potential in PE carry bags. The following microbes have also shown plastic degradation, including *Pseudomonas aeruginosa, Streptococcus aureus* B-324, *Aspergillus glaucus, Micrococcus lylae* B-429, *Aspergillus niger*, and *Pleurotus ostretus* (Aswale, 2010). They also showed that the *Phanerochaete chrysosporium* and *Pseudomonas aeruginosa* had the extreme biodegradation potency for the polyethylene (PE) carry bag samples. The rate of biodegradation reported was 50% by *Phanerochaete chrysosporium* and 35% for *Pseudomonas aeruginosa,* under the experimental conditions of 24°C at pH 4.

Priyanka and Archana (2011) studied plastic cups and polythene bags biodegradation potential of microbes isolated from various soil sources, including agricultural, sludge zone, energy park, sewage water, medicinal garden etc. The maximum biodegradation rate was shown by the fungal strains, *Aspergillus niger* and *Streptococcus lactis* bacteria. Other microbes, such as *Pseudomonas, Aspergillus nidulance, Bacillus subtilis, Proteus vulgaris, Staphylococcus aureus, Aspergillus glaucus, Aspergillus flavus, Penicillium,* and *Micrococcus luteus* have also shown the potency for degrading the plastic cups and polythene bags. The bacterial strain, *Brevibaccillus borstelensis* 707, has been reported for its active degradation of branched low-density polyethylene. In a study among 250 strains, *Trichoderma viride* was screened with active bioremediation potential. In this study, the fungal strain was incubated for a period of 60 days in municipal plastic waste and an average loss of weight (20%) was used for calculating the bioremediation rate (Venkatesh et al., 2021). Another study to degrade the polythene carry bags and cups was conducted with microbial consortia (Reddy, 2008). The microbial consortia included the fungal strains *Aspergillus nidulans, A. flavus, A. candidus, A. cremeus, A. niger* and *A. glaucus* and bacterial strains such as *Bacillus* sp., *Pseudomonas* sp., *Staphylococcus* sp., *Diplococcus* sp., *Streptococcus* sp., Micrococcus sp. and *Moraxella* sp., that were isolated from the soil. In the presence of mixed microbial consortia, a maximum biodegradation rate of about 12% weight loss was recorded in LDPE and a maximum reduction in tensile with an incubation period of 12 months was recorded for the high-density polythene (Reddy, 2008). In a study, *Pseudomonas* sp. was used to evaluate the degradation potential on the natural polythene with 6% starch and artificial polyethylene. About 46% and 29% of weight loss were recorded for natural polythene and artificial polythene, respectively (Nanda et al., 2010). *Aspergillus oryzae* degraded the HDPE films (0.1 lm width) at a rate of 72% after 90 days of incubation (Konduri et al., 2010). In another study, samples were collected from polythene dumping sites and incubated with the selected microbe for about 30 days and analyzed for weight loss and crystallinity percentage. Commercially available HDPE was actively biodegraded by *Pseudomonas* sp. and the *Arthrobacter* species and their biodegradation capability was found to be 15% and 12% respectively (Balasubramanian et al., 2010).

Biodegradation test on powdered low-density polythene was done using the fungal strains, *Penicillium pinophilum* and *Aspergillus niger*. With the 31 months of incubation, crystallinity, and percentage change in glassy were recorded. *A. niger* resulted in a maximum rate of 5% reduction in crystallinity and *Pencillium pinophilum* showed 11% of the change in glassy were recorded. The samples were incubated under the presence and absence of ethanol. Mineralization was higher for *Pencillium pinophilum* when incubated with ethanol than absence of ethanol (Volke-Sepulveda et al., 2002). In other study, LDPE powder was exposed to the biodegradation assay by using different microbial isolates (i.e., actinomycetes, fungi and bacteria) isolated from plastic waste dumping sites. Weight loss % was measured to calculate the biodegradation rate. After an incubation period of 6 months, the rate of biodegradation in weight loss was found to be 46.16%, 37% and 21% by *Streptomyces* KU8, *Aspergillus flavus* and *Pseudomonas* sp. respectively (Usha et al., 2011).

Polyurethane is susceptible to biodegradation despite its xenobiotic origins (Howard et al., 1999). A bacterial strain *Comamonas acidovorans* TB-35 has the polyester polyurethane (PUR)-degrading enzyme, PUR esterase, that can degrade Polyurethane (Akutsu et al., 1998). Biodegradation of polyurethane analyzed with the help of electron microscope by an increase in the number and size of pores on its surface (Cregut et al., 2013). Polyurethane-degrading microorganisms, including *Fusarium solani*, *Curvularia senegalensis*, *Aureobasidium pullulans* and *Cladosporidium* sp., were isolated (Crabbe et al., 1994). Some bacteria were also claimed to be capable of degrading polyurethane and they are strains of *Acinetobacter calcoaceticus*, *Arthrobacter globiformis*, *Pseudomonas aeruginosa*, *Pseudomonas cepacia*, *Pseudomonas putida*, and two other *Pseudomonas*-like species (El-Sayed et al., 1996). In addition, Stern and Howard (2000) reported that *Pseudomonas chlororaphis* encoded a lipase responsible for the degradation.

Plastic degradation can be further classified according to the agency causing it as follows:

a) Biodegradation - this is a natural process by which microorganism (bio-agents) breakdown plastics into smaller units in order to use the carbon and energy sources for growth and energy. This process can occur through hydro-biodegradation i.e., degradation due to hydrolytic cleavage of macromolecules and oxo-biodegradation i.e., degradation due to oxidation and cell-mediation;

b) Photo-degradation - this is brought about by the action of sunlight. It can result in an alteration in the physical and chemical structures of plastic, usually because of exposure to natural sunlight;

c) Thermo-oxidative degradation – this is caused by the slow oxidative breakdown at moderate temperatures;

d) Thermo-degradation caused by the action of high temperatures; and

e) Hydrolysis – this is the reaction of plastic with water.

Table 5 shows previous work on types of plastics, types of microbes and their effectiveness in plastic degradation. Generally, natural degradation of plastics does not occur in isolation. It occurs as a sequence of events starting with photo-degradation to thermo-oxidative degradation. Ultraviolet light generated by sunlight can also play a major role by providing the necessary energy required to incorporate oxygen atoms into the plastic (Andrady, 2011). This causes the plastic to become brittle and start to break into smaller units that can be metabolised by microorganisms. The microorganisms then convert the carbon in the plastic to carbon dioxide and water.

Challenges of Plastic Degradation in Compost

The potential for microbial degradation has been explored and established by different researchers. However, there are numerous underlining challenges associated with the biodegradation of plastics, especially in compost.

Table 5. Microbial degradation of plastics from different sources

Type of plastics	Source	Microorganism	Effects	Reference
Polyethylene	Waste management landfills	*Bacillus sp and Paenibacillus sp*	Reduced the dry weight by 14.7% after 60 days of incubation.	Park and Kim (2019)
	Marine	*Agios onoufrios and Kalathas*	Reduced weight by 19% after 6 months of incubation	Syranidou et al. (2017)
	Soil samples	*Aspergillus niger and Streptococcus lactis*	12.25% and 12.5% of biodegradation rate in 31 days incubation respectively.	Priyanka and Archana, 2011
	Mangrove sediment	*Bacillus gottheilii*	Reduced weight by 6.2% after 40 days of incubation	Auta et al. (2017)
	Compost	*Chelatococcus sp. E1*	Biodegradability reached 44.5% after 80 days of incubation at 58°C.	Jeon and Kim (2013)
	Dumping areas	*Pseudomonas sp.* and *Bacillus cereus*	12.5% of weight loss three months 20.10% of weight loss in 60 days	Aswale and Ade (2009)
	Municipal solid waste	*Trichoderma viride*		Venkatesh et al., 2021
Polypropylene	Municipal compost waste	*Bacillus sp.*	Reduced weight by 10-12% after 15 days of incubation at 37°C	Jain et al. (2018)
	Mangrove sediment	*Rhodococcus sp*	Reduced weight by 6.4% after 40 days of incubation	Auta et al. (2018)
	Mangrove sediment	*Bacillus gottheilii*	Reduced weight by 3.6% after 40 days of incubation	Auta et al. (2017

Table 5. (Continued)

Type of plastics	Source	Microorganism	Effects	Reference
LDPE, HDPE and polypropylene	Waste management landfills	*Aneurinibacillus aneurinilyticus* btDSCE01, *Brevibacillus agri* btDSCE02, *Brevibacillus sp.* btDSCE03 and *Brevibacillus brevis* btDSCE0	Reduced the weight of LDPE, HDPE and up to 58% after a period of 140 days at 50 °C.	Skariyachan et al. (2018)
LDPE	Plastic waste dumping sites	*Streptomyces* KU8 *Aspergillus flavus Pseudomonas* sp.	Reduced the weight by 46.16%, 37.09% and 20.63% respectively. with 6 months of incubation	Usha et al., 2011
HDPE and LDPE	Soil bed	Bacillus cereus, Bacillus pumilus and Arthrobacter	Reduced weight by 22.4% after 14 days of incubation	Satlewal et al. (2008)
Polystyrene	Mangrove sediment	*Bacillus gottheilii*	Reduced weight by 5.8% after 40 days of incubation	Auta et al. (2017)
Polyethylene terephthalate	Mangrove sediment	*Bacillus gottheilii*	Reduced weight by 3% after 40 days of incubation	Auta et al. (2017
Lignocellulose	unknown	*S. viridosporus T7A, S. badius 252 and S. setonii 75Vi2 (bacteria) and Phanerochaete chrysosporium (fungus)*	*50% reduction in tensile strength*	*Lee et al. (1991)*
Low branched (0.9 g/cm³) polyethylene	Soil	*Brevibacillus borstelensis strain 707*	*11% (gravimetric) and 30% (molecular) weights loss was reported at 50ºC after 30 days*	Hadad et al. (2005)
LPDE in the powdered form	Sea water	*Aspergillus versicolor and Aspergillus sp.*	*Maximum 4.16 g/L of CO₂ was released after degradation of the polythene*	*Pramla and Ramesh (2011)*

Type of plastics	Source	Microorganism	Effects	Reference
Low density polyethylene blended with starch	Lab culture	*Aspergillus niger, Penicillium funiculosum, Chaetomium globosum, Gliocladium virens* and *Pullularia pullulans*	Starch content was found directly proportional to the rate of degradation	Gilan et al. (2004)
Polyethylene bags	Landfill	*Serretia marscence*	22.2% of polythene degradation per month was recorded at pH 4, room temperature with regular shaking	*Aswale and Ade (2009)*
Polyethylene carry bags and plastics cups	Naturally buried polyethylene carry bags and cups in municipal composite	The predominant bacteria were *Bacillus sp., Staphylococcus sp., Streptococcus sp., Diplococcus sp., Micrococcus sp., Pseudomonas sp.* and *Moraxella* sp.; the redominant fungi were *Aspergillus niger. A. ornatus, A. nidulans, A. cremeus, A. flavus, A. candidus* and *A. glaucus*	In compost culture, highest percentage of weight loss (11.54%) was recorded in LDPE after 12 months whereas highest percent loss in tensile strength was reported with HDPE in the same time of incubation	*Reddy (2008)*
LDPE and LLDPE	Farm	*Bacillus cereus, B. megaterium, B. subtilis* and *Brevibacillus borstelensis*	Polythene films 75-85% (containing Fe stearate) and 31-67% (containing Ca stearate) at 45°C led to reduction in carbonyl index	*Abrusci et al. (2011)*
Branched low density polyethylene	Soil and mulch	*R. ruber* C208	8% of polyethylene degradation in 4 weeks	*Chandra and Rustgi (1997)*
HDPE and LDPE	Soil	*Bacillus, Micrococcus, Listeria* and *Vibrio*	Nearly 5% of weight loss after 8 weeks	*Kumar et al. (2007)*
Polyethylene bag wastes (pure water sachets)	Landfill	*Pseudomonas aeruginosa, Pseudomonas putida, Bacillus subtilis* and *Aspergillus niger*	After 8 weeks, only 1.2% weight loss was recorded when treated with 0.5 M HNO_3	*Nwachukwu et al. (2010)*

Some of these challenges are:

1) The volume of plastics generated per day globally in relation to the number of microorganisms that will be needed to break them down is a major challenge;

2) The environmental condition of each region/location differs. Hence, there is a need to investigate the native microbial consortium that will be responsible for biodegradation in the region under consideration. This process will likely involve significant time, energy and resources;

3) Previous findings have established that there are different microorganisms responsible for the biodegradation of different types of plastic. Therefore, for effective biodegradation of plastic within a compost pile, there must be the right microorganisms and right types of plastic in the pile;

4) The composting process is another challenge, as this varies -by composting facility and involves different types of microorganisms, e.g., aerobic and anaerobic composting;

5) The treatment of the compost feedstock with the identified and isolated microbes in a commercial setting can pose a major challenge. For instance, the identified microbes may have an adverse effect on human health when the compost is treated with such microbes;

6) Another major challenge of plastics decomposition is that during the process of composting, different microorganisms are present at different composting stages. Also, the structural composition and species distribution are probably most significantly affected by temperature distribution within the compost mix. For example, at curing stage or mesophilic phase microbes form the pioneer community, which rapidly breakdown complex chemical compounds in the materials resulting in the high temperature within the compost and thus, pave way for thermophilic microbes at above 45°C. The high temperature accelerates the breakdown of proteins, fats and complex carbohydrates like cellulose and hemicellulose; and

7) The heat can also affect the physical and chemical structure of plastics in the compost. The last is the mesophilic phase during which microorganisms again take over for the final phase of 'curing' or maturation of the remaining organic matter.

In brief summary, compost recipes, preparation methods and composting time play a significant role in shaping compost microbiota, which can also affect the ability of microorganisms to degrade plastics.

Conclusion

Based on literature information, it can be concluded that plastic materials are very useful items in our day-to-day life due to their versatility, wide range of usage and durability. However, the major constraint is sustainable plastic waste disposal through efficient degradation and diversion from landfills. Plastic waste, if not properly disposed of, accumulates in the environment and, in the long run, can end up in finished compost. This creates serious health, environmental and agro-ecological systems. Fortunately, compost is believed to house many microorganisms, some of which can break down plastics or have potential to break down plastics. The literature revealed many studies that have confirmed the use of different techniques which identified potential plastic degrading microorganisms using a variety of assessment methods. This microbial potential to breakdown plastics in compost will require comprehensive regional and local research to identify native microorganisms. It will also be important to investigate the residual effect of plastic degradation, if any. Currently, there is on-going research that is examining ways of evaluating the diversity of microbial communities in compost and the effect of the presence of plastics, using a second-generation metagenomics tool.

References

Abrusci, Concepción, Jesús Luís Pablos, Teresa Corrales, Josefa López-Marín, Irma Marín, and Fernando Catalina. 2011. "Biodegradation of photo-degraded mulching films based on polyethylenes and stearates of calcium and iron as pro-oxidant additives." *International Biodeterioration & Biodegradation* 65 (3):451-459.

Agathokleous, Evgenios, Ivo Iavicoli, Damià Barceló, and Edward J Calabrese. 2021. "Ecological risks in a 'plastic'world: a threat to biological diversity?" *Journal of Hazardous Materials* 417:126035.

Akutsu, Yukie, Toshiaki Nakajima-Kambe, Nobuhiko Nomura, and Tadaatsu Nakahara. 1998. "Purification and properties of a polyester polyurethane-degrading enzyme from Comamonas acidovorans TB-35." *Applied and Environmental Microbiology* 64 (1):62-67.

Al Hosni, Asma S, Jon K Pittman, and Geoffrey D Robson. 2019. "Microbial degradation of four biodegradable polymers in soil and compost demonstrating polycaprolactone as an ideal compostable plastic." *Waste Management* 97:105-114.

Alagha, Danah I, John N Hahladakis, Sami Sayadi, and Mohammad A Al-Ghouti. 2022. "Material flow analysis of plastic waste in the gulf co-operation countries (GCC) and the Arabian gulf: Focusing on Qatar." *Science of The Total Environment* 830:154745.

Alam, Pervez, and Kafeel Ahmade. 2013. "Impact of solid waste on health and the environment." *International Journal of Sustainable Development and Green Economics (IJSDGE)* 2 (1):165-168.

Alauddin, MB, Choudhury, E Baradie, and MSJ Hashmi. 1995. "Plastics and their machining: a review." *Journal of Materials Processing Technology* 54 (1-4):40-46.

Aldas, Miguel, Andrea Paladines, Vladimir Valle, Miguel Pazmiño, and Francisco Quiroz. 2018. "Effect of the prodegradant-additive plastics incorporated on the polyethylene recycling." *International Journal of Polymer Science* 2018.

Ali, Sameh Samir, Tamer Elsamahy, Eleni Koutra, Michael Kornaros, Mostafa El-Sheekh, Esraa A Abdelkarim, Daochen Zhu, and Jianzhong Sun. 2021. "Degradation of conventional plastic wastes in the environment: A review on current status of knowledge and future perspectives of disposal." *Science of The Total Environment* 771:144719.

Andrady, Anthony L. 2011. "Microplastics in the marine environment." *Marine pollution bulletin* 62 (8):1596-1605.

Antunes, Luciana Principal, Layla Farage Martins, Roberta Verciano Pereira, Andrew Maltez Thomas, Deibs Barbosa, Leandro Nascimento Lemos, Gianluca Major Machado Silva, Livia Maria Silva Moura, George Willian Condomitti Epamino, and Luciano Antonio Digiampietri. 2016. "Microbial community structure and dynamics in thermophilic composting viewed through metagenomics and metatranscriptomics." *Scientific reports* 6 (1):1-13.

Artham, Trishul, and Mukesh Doble. 2008. "Biodegradation of aliphatic and aromatic polycarbonates." *Macromolecular Bioscience* 8 (1):14-24.

Aswale, P.N 2010. "*Studies on bio-degradation of polythene.*" PhD diss., Dr Babasaheb Ambedkar Marathwada University, Aurangabad, India.

Aswale, PN, and AB Ade. 2009. "Effect of pH on biodegradation of polythene by Serretia marscence." *The Ecotech* 1:152-153.

Auta, Helen Shnada, Chijioke Uche Emenike, B Jayanthi, and Shahul Hamid Fauziah. 2018. "Growth kinetics and biodeterioration of polypropylene microplastics by Bacillus sp. and Rhodococcus sp. isolated from mangrove sediment." *Marine Pollution Bulletin* 127:15-21.

Auta, HS, CU Emenike, and SH Fauziah. 2017. "Screening of Bacillus strains isolated from mangrove ecosystems in Peninsular Malaysia for microplastic degradation." *Environmental Pollution* 231:1552-1559.

Balasubramanian, V, K Natarajan, B Hemambika, N Ramesh, CS Sumathi, R Kottaimuthu, and V Rajesh Kannan. 2010. "High-density polyethylene (HDPE)-degrading potential bacteria from marine ecosystem of Gulf of Mannar, India." *Letters in applied microbiology* 51 (2):205-211.

Barlaz, Morton A, Robert K Ham, and Daniel M Schaefer. 1989. "Mass-balance analysis of anaerobically decomposed refuse." *Journal of Environmental Engineering* 115 (6): 1088-1102.

Bläsing, Melanie, and Wulf Amelung. 2018. "Plastics in soil: Analytical methods and possible sources." *Science of the total environment* 612:422-435.

Boots, Bas, Connor William Russell, and Dannielle Senga Green. 2019. "Effects of microplastics in soil ecosystems: above and below ground." *Environmental science & technology* 53 (19):11496-11506.

Brachner, Andreas, Despina Fragouli, Iola F Duarte, Patricia MA Farias, Sofia Dembski, Manosij Ghosh, Ivan Barisic, Daniela Zdzieblo, Jeroen Vanoirbeek, and Philipp Schwabl. 2020. "Assessment of human health risks posed by nano-and microplastics is currently not feasible." *International Journal of Environmental Research and Public Health* 17 (23):8832.

Bradney, Lauren, Hasintha Wijesekara, Kumuduni Niroshika Palansooriya, Nadeeka Obadamudalige, Nanthi S Bolan, Yong Sik Ok, Jörg Rinklebe, Ki-Hyun Kim, and MB Kirkham. 2019. "Particulate plastics as a vector for toxic trace-element uptake by aquatic and terrestrial organisms and human health risk." *Environment international* 131:104937.

Braun, Melanie, Matthias Mail, Rene Heyse, and Wulf Amelung. 2021. "Plastic in compost: Prevalence and potential input into agricultural and horticultural soils." *Science of The Total Environment* 760:143335.

Brebu, Mihai. 2020. "Environmental degradation of plastic composites with natural fillers—a review." *Polymers* 12 (1):166.

Brinton, Will, Cyndra Dietz, Alycia Bouyounan, and Dan Matsch. 2018. "*Microplastics in Compost: The Environmental Hazards Inherent in the Composting of Plastic-Coated Paper Products.*"

Büks, Frederick, and Martin Kaupenjohann. 2020. "Global concentrations of microplastics in soils–a review." *Soil* 6 (2):649-662.

Campanale, Claudia, Carmine Massarelli, Ilaria Savino, Vito Locaputo, and Vito Felice Uricchio. 2020. "A detailed review study on potential effects of microplastics and additives of concern on human health." *International journal of environmental research and public health* 17 (4):1212.

Capitalist, Visual. 2018. "*Daily municipal solid waste generation per capita worldwide in 2018.*" accessed 14/07/2022. https://www.statista.com/statistics/689809/per-capital-msw-generation-by-country-worldwide/.

Chalmin, Philippe. 2019. "The history of plastics: from the Capitol to the Tarpeian Rock." *Field actions science reports. The Journal of Field Actions* (Special Issue 19):6-11.

Chamas, Ali, Hyunjin Moon, Jiajia Zheng, Yang Qiu, Tarnuma Tabassum, Jun Hee Jang, Mahdi Abu-Omar, Susannah L Scott, and Sangwon Suh. 2020. "Degradation rates of plastics in the environment." *ACS Sustainable Chemistry & Engineering* 8 (9):3494-3511.

Chandra, Shekhar, SS Kashyap, and A Singh. 2010. "Dengue syndrome: an emerging zoonotic disease." *North-East Veterinarian* 9 (4):21-22.

Chen, Chunzhao, Ling Chen, Ying Yao, Francisco Artigas, Qinghui Huang, and Wen Zhang. 2019. "Organotin release from polyvinyl chloride microplastics and concurrent photodegradation in water: Impacts from salinity, dissolved organic matter, and light exposure." *Environmental science & technology* 53 (18):10741-10752.

Corradini, Fabio, Pablo Meza, Raúl Eguiluz, Francisco Casado, Esperanza Huerta-Lwanga, and Violette Geissen. 2019. "Evidence of microplastic accumulation in agricultural soils from sewage sludge disposal." *Science of the total environment* 671:411-420.

Crabbe, Joel R, James R Campbell, Laura Thompson, Stephen L Walz, and Warren W Schultz. 1994. "Biodegradation of a colloidal ester-based polyurethane by soil fungi." *International Biodeterioration & Biodegradation* 33 (2):103-113.

Cregut, Mickael, M Bedas, M-J Durand, and Gerald Thouand. 2013. "New insights into polyurethane biodegradation and realistic prospects for the development of a sustainable waste recycling process." *Biotechnology advances* 31 (8):1634-1647.

Danso, Dominik, Christel Schmeisser, Jennifer Chow, Wolfgang Zimmermann, Ren Wei, Christian Leggewie, Xiangzhen Li, Terry Hazen, and Wolfgang R Streit. 2018. "New insights into the function and global distribution of polyethylene terephthalate (PET)-degrading bacteria and enzymes in marine and terrestrial metagenomes." *Applied and environmental microbiology* 84 (8):e02773-17.

De-la-Torre, Gabriel Enrique. 2020. "Microplastics: an emerging threat to food security and human health." *Journal of food science and technology* 57 (5):1601-1608.

Devi, Rajendran Sangeetha, Velu Rajesh Kannan, Krishnan Natarajan, Duraisamy Nivas, Kanthaiah Kannan, Sekar Chandru, and Arokiaswamy Robert Antony. 2016. "The role of microbes in plastic degradation." *Environ. Waste Manage* 341:341-370.

Dierkes, Georg, Tim Lauschke, Susanne Becher, Heike Schumacher, Corinna Földi, and Thomas Ternes. 2019. "Quantification of microplastics in environmental samples via pressurized liquid extraction and pyrolysis-gas chromatography." *Analytical and bioanalytical chemistry* 411 (26):6959-6968.

Ding, Ling, Zhuozhi Ouyang, Peng Liu, Tiecheng Wang, Hanzhong Jia, and Xuetao Guo. 2022. "Photodegradation of microplastics mediated by different types of soil: The effect of soil components." *Science of The Total Environment* 802:149840.

Dong, Youming, Minling Gao, Zhengguo Song, and Weiwen Qiu. 2020. "Microplastic particles increase arsenic toxicity to rice seedlings." *Environmental Pollution* 259: 113892.

Eartheasy. 2014. "*Solution for sustainable living.*" http://eartheasy.com/yard-garden/ composting.

Earthtalk. 2017. "*Recycling Different Plastics.*" accessed 14/07/2022. https://www. thoughtco.com/recycling-different-types-of-plastic-1203667.

El-Sayed, A Halim MM, Wafaa M Mahmoud, Edward M Davis, and Robert W Coughlin. 1996. "Biodegradation of polyurethane coatings by hydrocarbon-degrading bacteria." *International biodeterioration & biodegradation* 37 (1-2):69-79.

Elahi, Amina, Dilara Abbas Bukhari, Saba Shamim, and Abdul Rehman. 2021. "Plastics degradation by microbes: A sustainable approach." *Journal of King Saud University-Science* 33 (6):101538.

Eriksen, Marcus, Laurent CM Lebreton, Henry S Carson, Martin Thiel, Charles J Moore, Jose C Borerro, Francois Galgani, Peter G Ryan, and Julia Reisser. 2014. "Plastic pollution in the world's oceans: more than 5 trillion plastic pieces weighing over 250,000 tons afloat at sea." *PloS one* 9 (12):e111913.

Fackelmann, Gloria, and Simone Sommer. 2019. "Microplastics and the gut microbiome: how chronically exposed species may suffer from gut dysbiosis." *Marine pollution bulletin* 143:193-203.

Fauziah, Shahul Hamid, and Pariatamby Agamuthu. 2009. "Sustainable household organic waste management via vermicomposting." *Malaysian Journal of Science* 28 (2):135-142.

Fierer, Noah. 2017. "Embracing the unknown: disentangling the complexities of the soil microbiome." *Nature Reviews Microbiology* 15 (10):579-590.

Folino, Adele, Aimilia Karageorgiou, Paolo S Calabrò, and Dimitrios Komilis. 2020. "Biodegradation of wasted bioplastics in natural and industrial environments: A review." *Sustainability* 12 (15):6030.

Friend D, and Smith M. 2017. "*The Science of Composting.*" accessed 14/07/2022. https://web.extension.illinois.edu/homecompost/science.cfm.

Fuller, Stephen, and Anil Gautam. 2016. "A procedure for measuring microplastics using pressurized fluid extraction." *Environmental science & technology* 50 (11):5774-5780.

Gajšt, Tamara, Tine Bizjak, Andreja Palatinus, Svitlana Liubartseva, and Andrej Kržan. 2016. "Sea surface microplastics in Slovenian part of the Northern Adriatic." *Marine pollution bulletin* 113 (1-2):392-399.

Gao, Minling, Yu Liu, and Zhengguo Song. 2019. "Effects of polyethylene microplastic on the phytotoxicity of di-n-butyl phthalate in lettuce (*Lactuca sativa* L. var. ramosa Hort)." *Chemosphere* 237:124482.

Gattringer, Clemens W. 2021. "The Economics of Marine Plastic Pollution." In *Oxford Research Encyclopedia of Environmental Science*.

Gewert, Berit, Merle M Plassmann, and Matthew MacLeod. 2015. "Pathways for degradation of plastic polymers floating in the marine environment." *Environmental science: processes & impacts* 17 (9):1513-1521.

Ghatge, Sunil, Youri Yang, Jae-Hyung Ahn, and Hor-Gil Hur. 2020. "Biodegradation of polyethylene: a brief review." *Applied Biological Chemistry* 63 (1):1-14.

Ghosh, Swapan Kumar, Sujoy Pal, and Sumanta Ray. 2013. "Study of microbes having potentiality for biodegradation of plastics." *Environmental Science and Pollution Research* 20 (7):4339-4355.

Gibb, Bruce C. 2019. *"Plastics are forever."* Nature Publishing Group.

Gilan, IHYS, Y Hadar, and A Sivan. 2004. "Colonization, biofilm formation and biodegradation of polyethylene by a strain of Rhodococcus ruber." *Applied Microbiology and Biotechnology* 65 (1):97-104.

Godiya, Chirag Batukbhai, Luis Augusto Martins Ruotolo, and Weiquan Cai. 2020. "Functional biobased hydrogels for the removal of aqueous hazardous pollutants: current status, challenges, and future perspectives." *Journal of Materials Chemistry A* 8 (41):21585-21612.

Golwala, Harmita, Xueyao Zhang, Syeed Md Iskander, and Adam L Smith. 2021. "Solid waste: An overlooked source of microplastics to the environment." *Science of the Total Environment* 769:144581.

Gu, JD. 2017. Biodegradability of plastics: the pitfalls. *Applied Environmental Biotechnology*, 2 (1), 59-61.

Gu, Ji-Dong. 2003. "Microbiological deterioration and degradation of synthetic polymeric materials: recent research advances." *International biodeterioration & biodegradation* 52 (2):69-91.

Guanzon, Yvette B, and Robert J Holmer. 1993. "Composting of organic wastes: A main component for successful integrated solid waste management in Philippine cities." *National Eco-Waste Multisectoral Conference* and Techno Fair at Pryce Plaza Hotel, Cagayan de Oro City, Philippines.

Hadad, D, S Geresh, and Alex Sivan. 2005. "Biodegradation of polyethylene by the thermophilic bacterium Brevibacillus borstelensis." *Journal of applied microbiology* 98 (5):1093-1100.

Hahladakis, John N, Costas A Velis, Roland Weber, Eleni Iacovidou, and Phil Purnell. 2018. "An overview of chemical additives present in plastics: Migration, release, fate and environmental impact during their use, disposal and recycling." *Journal of hazardous materials* 344:179-199.

Halsband, Claudia, and Dorte Herzke. 2019. "Plastic litter in the European Arctic: what do we know?" *Emerging Contaminants* 5:308-318.

Hao, Zisu, Mei Sun, Joel J Ducoste, Craig H Benson, Scott Luettich, Marco J Castaldi, and Morton A Barlaz. 2017. "Heat generation and accumulation in municipal solid waste landfills." *Environmental science & technology* 51 (21):12434-12442.

Hauser, Russ, and AM Calafat. 2005. "Phthalates and human health." *Occupational and environmental medicine* 62 (11):806-818.

Henry, Asegun. 2014. "Thermal transport in polymers." *Annual review of heat transfer* 17.

Hoornweg, Daniel, Laura Thomas, and Lambert Otten. 1999. "Composting and its applicability in developing countries." *World Bank working paper series* 8:1-46.

Howard, Gary T, Carmen Ruiz, and Newton P Hilliard. 1999. "Growth of Pseudomonas chlororaphis on apolyester–polyurethane and the purification andcharacterization of a polyurethanase–esterase enzyme." *International biodeterioration & biodegradation* 43 (1-2):7-12.

Hu, Can, Bing Lu, Wensong Guo, Xiuying Tang, Xufeng Wang, Yinghao Xue, Long Wang, and Xiaowei He. 2021. "Distribution of microplastics in mulched soil in Xinjiang, China." *International Journal of Agricultural and Biological Engineering* 14 (2):196-204.

Hu, Qi, Xianyue Li, José M Gonçalves, Haibin Shi, Tong Tian, and Ning Chen. 2020. "Effects of residual plastic-film mulch on field corn growth and productivity." *Science of the Total Environment* 729:138901.

Huerta Lwanga, Esperanza, Hennie Gertsen, Harm Gooren, Piet Peters, Tamás Salánki, Martine Van Der Ploeg, Ellen Besseling, Albert A Koelmans, and Violette Geissen. 2016. "Microplastics in the terrestrial ecosystem: implications for Lumbricus terrestris (Oligochaeta, Lumbricidae)." *Environmental science & technology* 50 (5):2685-2691.

Igalavithana, Avanthi Deshani, Mahagama Gedara YL Mahagamage, Pradeep Gajanayake, Amila Abeynayaka, Premakumara Jagath Dickella Gamaralalage, Masataka Ohgaki, Miyuki Takenaka, Takayuki Fukai, and Norihiro Itsubo. 2022. "Microplastics and Potentially Toxic Elements: Potential Human Exposure Pathways through Agricultural Lands and Policy Based Countermeasures." *Microplastics* 1 (1):102-120.

Iiyoshi, Yuka, Yuji Tsutsumi, and Tomoaki Nishida. 1998. "Polyethylene degradation by lignin-degrading fungi and manganese peroxidase." *Journal of wood science* 44 (3):222-229.

Jain, Kimi, H Bhunia, and M Sudhakara Reddy. 2018. "Degradation of polypropylene–poly-L-lactide blend by bacteria isolated from compost." *Bioremediation Journal* 22 (3-4):73-90.

Jayaprakash, Vijayasree, and Uma Maheswari Devi Palempalli. 2018. "Effect of palmitic acid in the acceleration of polyethylene biodegradation by Aspergillus oryzae." *Journal of Pure and Applied Microbiology* 12 (4):2259-2269.

Jeffries, Peter, Silvio Gianinazzi, Silvia Perotto, Katarzyna Turnau, and José-Miguel Barea. 2003. "The contribution of arbuscular mycorrhizal fungi in sustainable maintenance of plant health and soil fertility." *Biology and fertility of soils* 37 (1):1-16.

Jenkins, MJ, and KL Harrison. 2008. "The effect of crystalline morphology on the degradation of polycaprolactone in a solution of phosphate buffer and lipase." *polymers for Advanced Technologies* 19 (12):1901-1906.

Jeon, Hyun Jeong, and Mal Nam Kim. 2013. "Isolation of a thermophilic bacterium capable of low-molecular-weight polyethylene degradation." *Biodegradation* 24 (1):89-98.

Jeyakumar, D, J Chirsteen, and Mukesh Doble. 2013. "Synergistic effects of pretreatment and blending on fungi mediated biodegradation of polypropylenes." *Bioresource technology* 148:78-85.

Kale, Gaurav, Thitisilp Kijchavengkul, Rafael Auras, Maria Rubino, Susan E Selke, and Sher Paul Singh. 2007. "Compostability of bioplastic packaging materials: an overview." *Macromolecular bioscience* 7 (3):255-277.

Kale, Swapnil Kisanrao, Amit G Deshmukh, Mahendra S Dudhare, and Vikram B Patil. 2015. "Microbial degradation of plastic: a review." *Journal of Biochemical Technology* 6 (2):952-961.

Karimifard, Shahab, and Mohammad Reza Alavi Moghaddam. 2018. "Application of response surface methodology in physicochemical removal of dyes from wastewater: a critical review." *Science of the Total Environment* 640:772-797.

Kijchavengkul, Thitisilp, and Rafael Auras. 2008. "Compostability of polymers." *Polymer International* 57 (6):793-804.

Kim, Hee-Soo, Han-Seung Yang, and Hyun-Joong Kim. 2005. "Biodegradability and mechanical properties of agro-flour–filled polybutylene succinate biocomposites." *Journal of Applied Polymer Science* 97 (4):1513-1521.

Kim, Yongho, Sumin Yeo, Joohee Kum, Hong-Gyu Song, and Hyoung T Choi. 2005. "Cloning of a manganese peroxidase cDNA gene repressed by manganese in Trametes versicolor." *Journal of Microbiology* 43 (6):569-571.

Kołodziejek, Jeremi. 2017. "Effect of seed position and soil nutrients on seed mass, germination and seedling growth in Peucedanum oreoselinum (Apiaceae)." *Scientific Reports* 7 (1):1-11.

Konduri, Mohan KR, Kuruganti S Anupam, Jakkula S Vivek, Rohini Kumar DB, and M Lakshmi Narasu. 2010. "Synergistic effect of chemical and photo treatment on the rate of biodegradation of high density polyethylene by indigenous fungal isolates." *International Journal of Biotechnology & Biochemistry* 6 (2):157-175.

Köninger, Julia, Panagiotis Panagos, Arwyn Jones, MJI Briones, and Alberto Orgiazzi. 2022. "In defence of soil biodiversity: Towards an inclusive protection in the European Union." *Biological Conservation* 268:109475.

Koschinsky, S, S Peters, F Schwieger, and CC Tebbe. 2000. "Applying molecular techniques to monitor microbial communities in composting processes." Progress in microbial ecology. *Proceedings of the International Symposium on Microbial Ecology*—8, in press.

Koutny, Marek, Pierre Amato, Marketa Muchova, Jan Ruzicka, and Anne-Marie Delort. 2009. "Soil bacterial strains able to grow on the surface of oxidized polyethylene film containing prooxidant additives." *International Biodeterioration & Biodegradation* 63 (3):354-357.

Kumar, Shristi, AAM Hatha, and KS Christi. 2007. "Diversity and effectiveness of tropical mangrove soil microflora on the degradation of polythene carry bags." *Revista de biología Tropical* 55 (3-4):777-786.

Lagalante, Anthony F, Courtney S Shedden, and Peter W Greenbacker. 2011. "Levels of polybrominated diphenyl ethers (PBDEs) in dust from personal automobiles in conjunction with studies on the photochemical degradation of decabromodiphenyl ether (BDE-209)." *Environment international* 37 (5):899-906.

Law, Kara Lavender. 2017. "Plastics in the marine environment." *Annual review of marine science* 9:205-229.

Leal Filho, Walter, Ulla Saari, Mariia Fedoruk, Arvo Iital, Harri Moora, Marija Klöga, and Viktoria Voronova. 2019. "An overview of the problems posed by plastic products and the role of extended producer responsibility in Europe." *Journal of cleaner production* 214:550-558.

Li, Boqing, Yunfei Ding, Xue Cheng, Dandan Sheng, Zheng Xu, Qianyu Rong, Yulong Wu, Huilin Zhao, Xiaofei Ji, and Ying Zhang. 2020. "Polyethylene microplastics affect the distribution of gut microbiota and inflammation development in mice." *Chemosphere* 244:125492.

Li, Penghui, Xiaodan Wang, Min Su, Xiaoyan Zou, Linlin Duan, and Hongwu Zhang. 2021. "Characteristics of plastic pollution in the environment: a review." *Bulletin of environmental contamination and toxicology* 107 (4):577-584.

Lian, Jiapan, Jiani Wu, Hongxia Xiong, Aurang Zeb, Tianzhi Yang, Xiangmiao Su, Lijuan Su, and Weitao Liu. 2020. "Impact of polystyrene nanoplastics (PSNPs) on seed germination and seedling growth of wheat (Triticum aestivum L.)." *Journal of hazardous materials* 385:121620.

Lim, Berlinda Kwee Hong, and Eng San Thian. 2021. "Biodegradation of polymers in managing plastic waste—A review." *Science of The Total Environment*: 151880.

Liu, Jiaoqin, Wenrui Xiang, Chenguang Li, Daniel J Van Hoomissen, Yumeng Qi, Nannan Wu, Gadah Al-Basher, Ruijuan Qu, and Zunyao Wang. 2020. "Kinetics and mechanism analysis for the photodegradation of PFOA on different solid particles." *Chemical Engineering Journal* 383:123115.

Liu, Jun, Ting Zhang, Sarah Piché-Choquette, Guofang Wang, and Jun Li. 2021. "Microplastic pollution in China, an invisible threat exacerbated by food delivery services." *Bulletin of Environmental Contamination and Toxicology* 107 (4):778-785.

Liwarska-Bizukojc, Ewa. 2021. "Effect of (bio) plastics on soil environment: A review." *Science of The Total Environment* 795:148889.

Ljung, Emelie, Kristina Borg Olesen, Per-Göran Andersson, Emma Fältström, Jes Vollertsen, Hans Bertil Wittgren, and Marinette Hagman. 2018. "Mikroplaster i kretsloppet." *Svenskt Vatten Utveckling Rapport* 13.

Luyt, Adriaan S, and Sarah S Malik. 2019. "Can biodegradable plastics solve plastic solid waste accumulation?" In *Plastics to energy*, 403-423. Elsevier.

MacLeod, Matthew, Hans Peter H Arp, Mine B Tekman, and Annika Jahnke. 2021. "The global threat from plastic pollution." *Science* 373 (6550):61-65.

Mayer, Melissa. 2018. *"What Is a Thermoplastic Polymer?,"* accessed 14/07/2022. https://sciencing.com/thermoplastic-polymer-5552849.html.

Mehmood, Ch Tahir, Ishtiaq A Qazi, Imran Hashmi, Samarth Bhargava, and Sriramulu Deepa. 2016. "Biodegradation of low density polyethylene (LDPE) modified with dye sensitized titania and starch blend using Stenotrophomonas pavanii." *International Biodeterioration & Biodegradation* 113:276-286.

Min, Kyungjun, Joseph D Cuiffi, and Robert T Mathers. 2020. "Ranking environmental degradation trends of plastic marine debris based on physical properties and molecular structure." *Nature communications* 11 (1):1-11.

Mistry M, Allaway D, Canepa P, and Rivin J. 2018. *"Comostable – How well does it predict the life cycle environmental impacts of packaging and food service ware?"* accessed 14/07/2022. https://www.oregon.gov/deq/FilterDocs/compostable.pdf.

Moharir, Rucha V, and Sunil Kumar. 2019. "Challenges associated with plastic waste disposal and allied microbial routes for its effective degradation: a comprehensive review." *Journal of Cleaner Production* 208:65-76.

Muenmee, Sutharat, Wilai Chiemchaisri, and Chart Chiemchaisri. 2015. "Microbial consortium involving biological methane oxidation in relation to the biodegradation of waste plastics in a solid waste disposal open dump site." *International Biodeterioration & Biodegradation* 102:172-181.

Müller, Rolf-Joachim. 2005. "Biodegradability of polymers: regulations and methods for testing." *Biopolymers Online: Biology Chemistry Biotechnology Applications* 10.

Nanda, Sonil, and Franco Berruti. 2021. "Municipal solid waste management and landfilling technologies: a review." *Environmental Chemistry Letters* 19 (2):1433-1456.

Nanda, Sonil, S Sahu, and Jayanthi Abraham. 2010. "Studies on the biodegradation of natural and synthetic polyethylene by Pseudomonas spp." *Journal of Applied Sciences and Environmental Management* 14 (2).

Neher, Deborah A, Thomas R Weicht, Scott T Bates, Jonathan W Leff, and Noah Fierer. 2013. "Changes in bacterial and fungal communities across compost recipes, preparation methods, and composting times." *PloS one* 8 (11):e79512.

Nwachukwu, Simon, Olayide Obidi, and Chinyere Odocha. 2010. "Occurrence and recalcitrance of polyethylene bag waste in Nigerian soils." *African Journal of Biotechnology* 9 (37):6096-6104.

OECD. 2022. *"Municipal solid waste recycling rates worldwide in 2020."* accessed 14/07/2022. https://www.statista.com/statistics/1052439/rate-of-msw-recycling-worldwide-by-key-country/.

Okubo, Nami, Shunichi Takahashi, and Yoshikatsu Nakano. 2018. "Microplastics disturb the anthozoan-algae symbiotic relationship." *Marine pollution bulletin* 135:83-89.

Olaosebikan Oluwatosin, O, Moses N Alo, Uchenna I Ugah, and M Olayemi Albert. 2014. *"Environmental Effect on Biodegradability of Plastic and Paper Bags."*

Ozores-Hampton, Monica, Thomas A Obreza, and George Hochmuth. 1998. "Using composted wastes on Florida vegetable crops." *HortTechnology* 8 (2):130-137.

Park, Seon Yeong, and Chang Gyun Kim. 2019. "Biodegradation of micro-polyethylene particles by bacterial colonization of a mixed microbial consortium isolated from a landfill site." *Chemosphere* 222:527-533.

Pathak, Vinay Mohan. 2017. "Review on the current status of polymer degradation: a microbial approach." *Bioresources and Bioprocessing* 4 (1):1-31.

Pérez-Piqueres, Ana, Véronique Edel-Hermann, Claude Alabouvette, and Christian Steinberg. 2006. "Response of soil microbial communities to compost amendments." *Soil Biology and Biochemistry* 38 (3):460-470.

Plastics Europe. 2021. *"Distribution of global plastic materials production in 2020."* accessed 14/07/2022. https://www.statista.com/statistics/281126/global-plastics-production-share-of-various-countries-and-regions/.

Platt, Brenda, Nora Goldstein, Craig Coker, and Sally Brown. 2014. "State of Composting in the US." *Institute for Local Self-Reliance*:1-131.

Population Matters. 2021. *"Biodiversity: Population Matters."* https://population matters.org/.

Pospíšil, Jan, and Stanislav Nešpůrek. 1997. *"Highlights in chemistry and physics of polymer stabilization."* Macromolecular Symposia.

Priyanka, Nayak, and Tiwari Archana. 2011. "Biodegradability of polythene and plastic by the help of microorganism: a way for brighter future." *J Environ Anal Toxicol* 1 (4):1000111.

Qi, Yueling, Xiaomei Yang, Amalia Mejia Pelaez, Esperanza Huerta Lwanga, Nicolas Beriot, Henny Gertsen, Paolina Garbeva, and Violette Geissen. 2018. "Macro-and micro-plastics in soil-plant system: effects of plastic mulch film residues on wheat (Triticum aestivum) growth." *Science of the Total Environment* 645:1048-1056.

Qu, Ruijuan, Chenguang Li, Xiaoxue Pan, Xiaolan Zeng, Jiaoqin Liu, Qingguo Huang, Jianfang Feng, and Zunyao Wang. 2017. "Solid surface-mediated photochemical transformation of decabromodiphenyl ether (BDE-209) in aqueous solution." *Water research* 125:114-122.

Raaman, N, N Rajitha, A Jayshree, and R Jegadeesh. 2012. "Biodegradation of plastic by Aspergillus spp. isolated from polythene polluted sites around Chennai." *J Acad Indus Res* 1 (6):313-316.

Ramachandra, TV, HA Bharath, Gouri Kulkarni, and Sun Sheng Han. 2018. "Municipal solid waste: Generation, composition and GHG emissions in Bangalore, India." *Renewable and Sustainable Energy Reviews* 82:1122-1136.

Raquez, Jean-Marie, Audrey Bourgeois, Heidi Jacobs, Philippe Degée, Michael Alexandre, and Philippe Dubois. 2011. "Oxidative degradations of oxodegradable LDPE enhanced with thermoplastic pea starch: Thermo-mechanical properties, morphology, and UV-ageing studies." *Journal of Applied Polymer Science* 122 (1):489-496.

Rawat, US, and NK Agarwal. 2015. "Biodiversity: Concept, threats and conservation." *Environment Conservation Journal* 16 (3):19-28.

Ray, Sudip, and Ralph P Cooney. 2018. "Thermal degradation of polymer and polymer composites." In *Handbook of environmental degradation of materials*, 185-206. Elsevier.

Raziyafathima, M, PK Praseetha, and IRS Rimal. 2016. "Microbial degradation of plastic waste: a review." *Chemical and Biological Sciences* 4:231-42.

Reddy, R Mallikarjuna. 2008. *"Impact of soil composting using municipal solid waste on biodegradation of plastics."*

Reusch, W. 2013. *"Polymers."* accessed 02/07/2022. https://www2.chemistry.msu.edu/faculty/reusch/virttxtjml/polymers.htm.

Ribitsch, Doris, Enrique Herrero Acero, Agnieszka Przylucka, Sabine Zitzenbacher, Annemarie Marold, Caroline Gamerith, Rupert Tscheließnig, Alois Jungbauer, Harald Rennhofer, and Helga Lichtenegger. 2015. "Enhanced cutinase-catalyzed hydrolysis of polyethylene terephthalate by covalent fusion to hydrophobins." *Applied and Environmental Microbiology* 81 (11):3586-3592.

Ritchie, Hannah. 2021. "Where does the plastic in our oceans come from." *Our World Data* 1.

Ruiz-Dueñas, Francisco J, and Ángel T Martínez. 2009. "Microbial degradation of lignin: how a bulky recalcitrant polymer is efficiently recycled in nature and how we can take advantage of this." *Microbial biotechnology* 2 (2):164-177.

Rujnić-Sokele, Maja, and Ana Pilipović. 2017. "Challenges and opportunities of biodegradable plastics: A mini review." *Waste Management & Research* 35 (2):132-140.

Santo, Miriam, Ronen Weitsman, and Alex Sivan. 2013. "The role of the copper-binding enzyme–laccase–in the biodegradation of polyethylene by the actinomycete Rhodococcus ruber." *International Biodeterioration & Biodegradation* 84:204-210.

Sardon, Haritz, and Andrew P Dove. 2018. "Plastics recycling with a difference." *Science* 360 (6387):380-381.

Satlewal, Alok, Ravindra Soni, MGH Zaidi, Yogesh Shouche, and Reeta Goel. 2008. "Comparative biodegradation of HDPE and LDPE using an indigenously developed microbial consortium." *Journal of microbiology and biotechnology* 18 (3):477-482.

Scheurer, Michael, and Moritz Bigalke. 2018. "Microplastics in Swiss floodplain soils." *Environmental science & technology* 52 (6):3591-3598.

Scopetani, Costanza, David Chelazzi, Alessandra Cincinelli, Tania Martellini, Ville Leiniö, and Jukka Pellinen. 2022. "Hazardous contaminants in plastics contained in compost and agricultural soil." *Chemosphere* 293:133645.

Sen, Sudip Kumar, and Sangeeta Raut. 2015. "Microbial degradation of low density polyethylene (LDPE): A review." *Journal of Environmental Chemical Engineering* 3 (1):462-473.

Shah, Aamer Ali, Fariha Hasan, Abdul Hameed, and Safia Ahmed. 2008. "Biological degradation of plastics: a comprehensive review." *Biotechnology advances* 26 (3):246-265.

Sharma, Marnika, Pratibha Sharma, Anima Sharma, and Subhash Chandra. 2015. "Microbial degradation of plastic-a brief review." *CIBTech Journal of Microbiology* 4 (1):85-89.

Singh, Baljit, and Nisha Sharma. 2008. "Mechanistic implications of plastic degradation." *Polymer degradation and stability* 93 (3):561-584.

Siracusa, Valentina. 2019. "Microbial degradation of synthetic biopolymers waste." *Polymers* 11 (6):1066.

Sivasankari, S, and T Vinotha. 2014. "In vitro degradation of plastics (plastic cup) using Micrococcus luteus and Masoniella Sp." *Sch. Acad. J. Biosci* 2 (2):85-89.

Skariyachan, Sinosh, Amulya A Patil, Apoorva Shankar, Meghna Manjunath, Nikhil Bachappanavar, and S Kiran. 2018. "Enhanced polymer degradation of polyethylene and polypropylene by novel thermophilic consortia of Brevibacillus sps. and Aneurinibacillus sp. screened from waste management landfills and sewage treatment plants." *Polymer Degradation and Stability* 149:52-68.

Spigolon, Luciana MG, Mariana Giannotti, Ana P Larocca, Mario AT Russo, and Natália da C Souza. 2018. "Landfill siting based on optimisation, multiple decision analysis, and geographic information system analyses." *Waste Management & Research* 36 (7):606-615.

Stern, Robert V, and Gary T Howard. 2000. "The polyester polyurethanase gene (pueA) from Pseudomonas chlororaphis encodes a lipase." *FEMS Microbiology Letters* 185 (2):163-168.

Stubbins, Aron, Kara Lavender Law, Samuel E Muñoz, Thomas S Bianchi, and Lixin Zhu. 2021. "Plastics in the Earth system." *Science* 373 (6550):51-55.

Syranidou, Evdokia, Katerina Karkanorachaki, Filippo Amorotti, Eftychia Repouskou, Kevin Kroll, Boris Kolvenbach, Philippe FX Corvini, Fabio Fava, and Nicolas Kalogerakis. 2017. "Development of tailored indigenous marine consortia for the degradation of naturally weathered polyethylene films." *PloS one* 12 (8):e0183984.

Teuten, Emma L, Jovita M Saquing, Detlef RU Knappe, Morton A Barlaz, Susanne Jonsson, Annika Björn, Steven J Rowland, Richard C Thompson, Tamara S Galloway, and Rei Yamashita. 2009. "Transport and release of chemicals from plastics to the environment and to wildlife." *Philosophical transactions of the royal society B: biological sciences* 364 (1526):2027-2045.

Thakur, Pooja. 2012. *"Screening of plastic degrading bacteria from dumped soil area."*

Thompson, Richard C, Charles J Moore, Frederick S Vom Saal, and Shanna H Swan. 2009. "Plastics, the environment and human health: current consensus and future trends." *Philosophical transactions of the royal society B: biological sciences* 364 (1526): 2153-2166.

Tian, Lili, Cheng Jinjin, Rong Ji, Yini Ma, and Xiangyang Yu. 2022. "Microplastics in agricultural soils: sources, effects, and their fate." *Current Opinion in Environmental Science & Health* 25:100311.

Tokiwa, Yutaka, Buenaventurada P Calabia, Charles U Ugwu, and Seiichi Aiba. 2009. "Biodegradability of plastics." *International journal of molecular sciences* 10 (9):3722-3742.

Tokiwa, Yutaka, and Tomoo Suzuki. 1978. "Hydrolysis of polyesters by Rhizopus delemar lipase." *Agricultural and Biological Chemistry* 42 (5):1071-1072.

Tong, Xuneng, Mui-Choo Jong, Jingjie Zhang, Luhua You, and Karina Yew-Hoong Gin. 2021. "Modelling the spatial and seasonal distribution, fate and transport of floating plastics in tropical coastal waters." *Journal of Hazardous Materials* 414:125502.

Toussaint, Brigitte, Barbara Raffael, Alexandre Angers-Loustau, Douglas Gilliland, Vikram Kestens, Mauro Petrillo, Iria M Rio-Echevarria, and Guy Van den Eede. 2019. "Review of micro-and nanoplastic contamination in the food chain." *Food Additives & Contaminants: Part A* 36 (5):639-673.

Tribedi, Prosun, Subhasis Sarkar, Koushik Mukherjee, and Alok K Sil. 2012. "Isolation of a novel Pseudomonas sp from soil that can efficiently degrade polyethylene succinate." *Environmental Science and Pollution Research* 19 (6):2115-2124.

USDA. 2010. *Environmental Engineering National Engineering Handbook.* .

Usha, Rajamanickam, T Sangeetha, and M Palaniswamy. 2011. "Screening of polyethylene degrading microorganisms from garbage soil." *Libyan Agric Res Cent J Int* 2 (4):200-4.

Venkatesh, S, Shahid Mahboob, Marimuthu Govindarajan, Khalid A Al-Ghanim, Zubair Ahmed, Norah Al-Mulhm, R Gayathri, and S Vijayalakshmi. 2021. "Microbial degradation of plastics: Sustainable approach to tackling environmental threats facing big cities of the future." *Journal of King Saud University-Science* 33 (3):101362.

Vivi, Viviane Karolina, Sandra Mara Martins-Franchetti, and Derlene Attili-Angelis. 2019. "Biodegradation of PCL and PVC: Chaetomium globosum (ATCC 16021) activity." *Folia microbiologica* 64 (1):1-7.

Volke-Sepúlveda, T, G Saucedo-Castañeda, M Gutiérrez-Rojas, A Manzur, and E Favela-Torres. 2002. "Thermally treated low density polyethylene biodegradation by Penicillium pinophilum and Aspergillus niger." *Journal of applied polymer science* 83 (2):305-314.

Wagg, Cameron, S Franz Bender, Franco Widmer, and Marcel GA Van Der Heijden. 2014. "Soil biodiversity and soil community composition determine ecosystem multifunctionality." *Proceedings of the National Academy of Sciences* 111 (14):5266-5270.

Walter, Rylie. 2019. *"Skin, Bones+ Bags: Investigating the Death of Marine Ecosystems."*

Wan, Yong, Chenxi Wu, Qiang Xue, and Xinminnan Hui. 2019. "Effects of plastic contamination on water evaporation and desiccation cracking in soil." *Science of the Total Environment* 654:576-582.

Wang, Jiao, Xianhua Liu, Yang Li, Trevor Powell, Xin Wang, Guangyi Wang, and Pingping Zhang. 2019. "Microplastics as contaminants in the soil environment: A mini-review." *Science of the total environment* 691:848-857.

Watt, Ethan, Maisyn Picard, Benjamin Maldonado, Mohamed A Abdelwahab, Deborah F Mielewski, Lawrence T Drzal, Manjusri Misra, and Amar K Mohanty. 2021. "Ocean plastics: environmental implications and potential routes for mitigation–a perspective." *RSC advances* 11 (35):21447-21462.

Wayman, Chloe, and Helge Niemann. 2021. "The fate of plastic in the ocean environment–a minireview." *Environmental Science: Processes & Impacts* 23 (2):198-212.

Webb, Hayden K, Jaimys Arnott, Russell J Crawford, and Elena P Ivanova. 2012. "Plastic degradation and its environmental implications with special reference to poly (ethylene terephthalate)." *Polymers* 5 (1):1-18.

Wei, Ren, and Nick Wierckx. 2021. "Microbial Degradation of Plastics." *Frontiers in Microbiology* 12:635621.

Weithmann, Nicolas, Julia N Möller, Martin GJ Löder, Sarah Piehl, Christian Laforsch, and Ruth Freitag. 2018. "Organic fertilizer as a vehicle for the entry of microplastic into the environment." *Science advances* 4 (4):eaap8060.

World Bank. 2020. "Solid Waste Management." accessed 14/07/2022. http://www.world bank.org/en/topic/urbandevelopment/brief/solid-waste-management.

Zhang, GS, and YF Liu. 2018. "The distribution of microplastics in soil aggregate fractions in southwestern China." *Science of the Total Environment* 642:12-20.

Zhang, Lishan, Yuanshan Xie, Junyong Liu, Shan Zhong, Yajie Qian, and Pin Gao. 2020. "An overlooked entry pathway of microplastics into agricultural soils from application of sludge-based fertilizers." *Environmental Science & Technology* 54 (7):4248-4255.

Zhang, Shaoliang, Xiaomei Yang, Hennie Gertsen, Piet Peters, Tamás Salánki, and Violette Geissen. 2018. "A simple method for the extraction and identification of light density microplastics from soil." *Science of the Total Environment* 616:1056-1065.

Zhang, Yalin, Xiaoting Zhang, Xinyu Li, and Defu He. 2022. "Interaction of microplastics and soil animals in agricultural ecosystems." *Current Opinion in Environmental Science & Health*:100327.

Zhao, Kai, Yunman Wei, Jianhong Dong, Penglu Zhao, Yuezhu Wang, Xinxiang Pan, and Junsheng Wang. 2021. "Separation and characterization of microplastic and nanoplastic particles in marine environment." *Environmental Pollution*:118773.

Zhou, Bin, Lixia Zhao, Yuebo Wang, Yang Sun, Xiaojing Li, Huijuan Xu, Liping Weng, Zheng Pan, Side Yang, and Xingping Chang. 2020. "Spatial distribution of phthalate esters and the associated response of enzyme activities and microbial community composition in typical plastic-shed vegetable soils in China." *Ecotoxicology and Environmental Safety* 195:110495.

Zhu, Bo-Kai, Yi-Meng Fang, Dong Zhu, Peter Christie, Xin Ke, and Yong-Guan Zhu. 2018. "Exposure to nanoplastics disturbs the gut microbiome in the soil oligochaete Enchytraeus crypticus." *Environmental Pollution* 239:408-415.

Zhu, Fengxiao, Changyin Zhu, Chao Wang, and Cheng Gu. 2019. "Occurrence and ecological impacts of microplastics in soil systems: a review." *Bulletin of environmental contamination and toxicology* 102 (6):741-749.

Chapter 2

Public Participation in the Implementation of Sustainable Solid Waste Management in Soweto, South Africa

Olusola Olaitan Ayeleru[1,2,*], PhD
Helen Uchenna Modekwe[2,3], PhD
Felix Ndubisi Okonta[4], PhD
Peter Apata Olubambi[1], PhD
and Freeman Ntuli[5], PhD

[1]Centre for Nanoengineering and Tribocorrosion (CNT), University of Johannesburg, Johannesburg, South Africa
[2]Conserve Africa Initiative, Osogbo, Osun State, Nigeria
[3]Department of Chemical Engineering, University of Johannesburg, South Africa
[4]Department of Civil Engineering, University of Johannesburg, South Africa
[5]Chemical, Materials and Metallurgical Engineering Department, Faculty of Engineering and Technology, Botswana International University of Science and Technology, Palapye, Botswana

Abstract

Inadequate management of waste is gradually turning out to be an issue of concern across the globe due to poor public participation. Participation of members of the public therefore becomes very crucial for a sustainable solid waste management to be achieved. The objective of this paper is to evaluate the perception of the public with regards to the current waste management system in Southwestern Township (Soweto), South Africa

* Corresponding Author's Emails: olusolaolt@gmail.com; olusola@conserveafricainitiative.org.

In: Municipal Solid Waste Management and Improvement Strategies
Editor: Adam Fitz
ISBN: 979-8-88697-720-2

as a way of getting the public involved in the implementation of a sustainable solid waste management (SSWM) or Zero waste (ZW) project. To achieve this, a questionnaire consisting of 48 questions was formulated and administered in four communities consisting of informal settlement, middle- and high-income areas. A total of 150 questionnaires were administered but only 118 was returned. Data were collected and analysed using SPSS software with 95% confidence level. Results showed that 51% of respondents were not satisfied with the services, 71% did not know where their collected waste was taken to, for final disposal and 77% of respondents did not know who to contact if they have issues with their waste collection services. From the overall analysis, it was concluded that the people are not properly educated on environmental matters, but they are willing to support Zero waste.

Keywords: zero waste, public participation, sustainable solid waste management, Soweto, Pikitup Johannesburg (Pty) Ltd, South Africa

Introduction

In developing countries (DCs), waste management (WM) is confronted with several issues (CSIR, 2011). To manage municipal solid waste (MSW) in urban centres is one of the greatest challenges in DCs owing to insufficient funding, rapid population growth and influx of people to urban centres (Pokhrel & Viraraghavan, 2005). Essentially, municipal solid waste management (MSWM) absorbs up to 1 per cent of Gross National Product (GNP) and 20 to 40 per cent of municipal revenues in DCs. Notwithstanding, when solid waste (SW) is managed well, it offers jobs to up to 6 workers for each 1,000 unemployed persons. This figure could identify with up to 2% of the national workforce. Generally, the service is not as frequently as should be expected under the circumstances because more than 50% of the refuse generated in urban groups is usually uncollected and other populated districts of the urban zones do not receive regular attention. Hence, poor communities remain underserviced by municipal solid waste management (MSWM) (Coffey & Coad, 2010). The challenges of MSWM in DCs are huge. In many urban areas in DCs, the waste collection vehicles break down frequently. Consequently, there is typically accumulation of wastes and in rural areas there is little or no service coverage. Rustic inhabitants effortlessly dump waste on any empty land, stream or even burn it, along these lines contaminating the air. There are no transfer stations in some DCs. Therefore, waste is dumped by the roadsides. Constrained spending plans are discharged to the districts; henceforth, no quality services can be guaranteed (Ogwueleka, 2009). These issues are both social and economic in nature. They run from low coverage, open dumping, open burning and dumping into the waterway. The effect of all of these on the wellbeing of the general population and the earth is extremely startling. A portion of the negative impacts are reproducing of flies which transmit diseases to the community members. Water bodies also get polluted and serve as source of water borne infections to members of the community. One of the economic facets of the issue of MSWM is lack of labour to discharge the SW services efficiently. Most officers charged with the responsibility are unprofessional in the field and there is also inadequate planning for solid waste management (SWM) at all levels (Hisashi & Kuala, 1997).

In South Africa (SA), management of municipal solid waste (MSW) is currently facing diverse issues. Some of the issues range from rapid economic and population growth; complex waste streams, little or no idea of waste flows and national waste balance, backlog of wastes in some settlement owing insufficient service coverage, policies, and regulations not favourable to WM, lack of recycling facilities, overstretching of old infrastructure, and handy WM methods are very costly compared to landfilling (DEA, 2011). Currently, the focus has been shifted from landfilling to recycling/circular economy (CE) despite recycling being quite expensive. Studies have shown that CE or Zero Waste (ZW) is the answer to the current waste management issues. CE is a concept relating to sustainable development (SD). CE focuses on ecological modernization, green growth and low carbon development which is ultimately leading to SD. To achieve SD in a country like SA, economic growth and environmental protection are important factors that must be put into consideration. For a CE or ZW to be a huge success, changes must occur in terms of culture, principle, and practice of the current WM methods. Also, active participation of all and sundry becomes very crucial. Aforesaid involvement includes the waste generators, waste pickers, recycling industries, waste collection contractors, waste management agencies, politicians, national government, public sectors, higher institution of learning and financial institutions (CIWM, 2014; Squires, 2006; Zhou et al., 2014).

This paper aims to evaluate the level at which people are conversant with their environment and the need to getting the public involved in the implementation of a sustainable solid waste management (SSWM).

Research Design and Methodology

The central goal of this research methodology was to find out how people are conversant with their environment and to identify the issues that relate to the collection and disposal of MSW in their communities. To achieve the research objectives, information was gathered from four communities through quantitative research methods. Quantitative data were gathered by means of a structured questionnaire survey conducted on the households within the chosen study areas (Bryman, 2004). The chosen communities include the informal settlement (low-income), middle-income and high-income classes. The communities are Naledi extension and Dobsonville estate. Secondary information was gathered through articles, reports, books, published and unpublished materials, and from the internets.

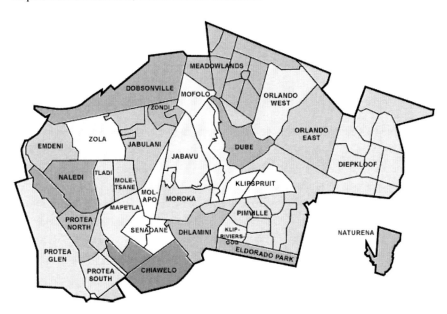

Figure 1. Map of Soweto.

Area of Study

South-western townships (Soweto) came into existence when African mine workers were forced from Brickfields to a "sanitary camp" on the farm Klipspruit after a plague broke-out in 1904. The township gradually emerged from there in 1963. It is located on 15 km south-west of the Johannesburg Central Business District (CBD) and bordered by the Westrand District Municipality in the west, N1 highway in the east, a mining belt in the north and the N12 in the south. Soweto (Figure 1) is the largest township in the whole of South Africa. Its estimated population is ~1.8 million and the population of women constituted about 51%. It occupies about 150 km^2 of the City of Johannesburg (CoJ). It home about 43% of the entire population of CoJ (CoJ, 2006 – 2011).

Primary Data Collection and Analysis (Quantitative)

Quantitative method of research measures the behaviour of consumers, their knowledge, and opinions. This attempts to provide answers to questions like how much, how often, how many, when, who and what; using a data collection instrument called questionnaire. A questionnaire is the instrument that is mostly used in research, and developing one is partly science and partly art (Cooper & Schindler, 2008). A structured questionnaire was administered to households in four communities. 150 questionnaires were distributed but only 118 were returned. The distributions of questionnaire to the sampled population were done by the Researcher alongside some staff of Pikitup Johannesburg (Pty) Ltd (the municipality); few were filled on the spot and some were returned after they had been completed. The respondents' identities were not requested. This made them feel relaxed; hence, they responded positively. Households were interviewed with the help of the questionnaire. Households' levels of awareness, attitudes, concerns, and willingness to participate in ZW project and other general issues of concern with respect to solid waste management (SWM) were evaluated using the questionnaire. Data were obtained from the households' survey using stratified random sampling technique. A distribution of sample was chosen so that a representative sample size and the heterogeneity of the study population could be represented. The data were analysed using SPSS statistics descriptive program.

Secondary Data Collection (Literature Review)

Secondary data was reviewed, and this has been very helpful in this research. Official reports, books, encyclopaedia, international standards, articles, legal documents, published and unpublished literature and case studies were consulted. The information extracted from all these resource materials cover a wide range of issues around MSWM and the ZW concept (Hox & Boeije, 2005). Advancement in technology has resulted in having data collected, compiled, and archived; and researcher are currently having access to them. The use of secondary data is becoming popular. In this study, the area of study and the research questions are the parameters that determined the method used by the researcher (Johnston, 2014).

Research Design and Sources of Data

The research method used for this work was the survey method. A survey is a method of research used to gather data from a group of people with the aid of methodized questionnaire. Questionnaire is just an aspect of survey. Survey also involves selection of population, pre-testing instruments, analysis of data etc. (Office of Planning and Institutional Assessment, 2006). Survey is mostly used to gather information from a small group from a population and the output is often generalized to the overall population (Snijkers et al., 2013). This was done with the belief that surveys are used to gather up to date information. The survey method was chosen in preference to other methods like content analysis because of its capacity to measure human attitudes and opinions. The content analysis is a research method used to analyse words or texts and draw up a conclusion on the messages. Researchers also used this method to carefully examine human interactions and to analyse characters on television (TV), films and novels (Neuendorf, 2002). The researcher used closed questions, open-ended questions, and contingency questions to generate authentic data from the respondents (Siniscalco & Auriat, 2015).

The survey questionnaire covered; ways households handle their solid waste (SW), households' level of awareness on WM, households' level of satisfaction with respect to the services offered to them by the municipality and their willingness to support recycling and ZW.

Limitations of the Study

Like any research, no matter how it is structured, it would usually have its limitation. Limitations are challenges encountered while conducting a study and they are usually beyond the control of the researcher (Simon & Goes,

2013). This study has several features that limit the generalization of its findings. Some of the challenges encountered were in the distribution of questionnaires and the collection of data owing to the attitudes of some of the respondents. Some of the respondents were very hostile and some questionnaires were not properly completed.

Results and Discussions

Household Identification

The demographics of the respondents and their households' identification are presented in Table 1. The groups comprised of Naledi extension informal settlement (NEIS) which represented the low-income class, Naledi extension RDP (Reconstruction and Development Programme) which represented a middle-income class, Dobsonville RDP also represented a middle-income class and Dobsonville Bond represented the high-income class. There were about 35% respondents from NEIS, 28% respondents from Naledi RDP, 18% respondents from Dobsonville Bond and 19% from Dobsonville RDP. About 51% of the respondents were male and 49% were female. This implies that there were more males than females among the respondents selected for this study. This is an indication that men are also interested in environmental matters, they are concerned about WM, and they are also looking forward to better waste collection services.

General and Personal Household Information

Some personal questions about the respondents and their family are presented in the seventh section of the questionnaire and the corresponding responses are as shown in Table 1.

From the sampled population, about 87% of the respondents said household members were the owners of their apartments, 4% said their apartments were owned by government and 9% said their apartments were owned by private owners. This shows that most of the population may not be paying house rents since their apartments were owned by members of their households. The study also showed the distribution of people living in their

Table 1. Demographics, households' identification, and general household information of respondents

Background identification	Respondent Groups			
Response codes	Husband	Wife	Others	No response
	39%	33%	27%	1%
Response from the Communities	Naledi extension informal settlement	Naledi extension RDP	Dobsonville RDP	Dobsonville Bond
	39%	28%	19%	18%
Gender	Male		Female	
	51%		49%	
Household information	Respondents			
Number of people living in a household	0	1	2	3	Others	No response
	3%	28%	23%	22%	24%	0%
Retired adults in a household	0	1	2	3	Others	No response
	0%	19%	5%	2%	33%	48%
Owners of the apartments	Household members		Government		Private owners	
	87%		4%		9%	
Pensioners living in a household	0	1	2	3	Others	No response
	20%	20%	2%	0%	0%	58%
Adults with regular income	0	1	2	3	4	No response
	8%	42%	18%	3%	1%	28%
Adults without seasonal income	0	1	2	3	4	No response
	15%	32%	21%	3%	1%	28%
Highest educational qualification	Never been to school	Primary school	Secondary school	High school	University	No response
	5%	30%	46%	13%	5%	1%

various households. 3% of the respondents said only 1 person lived in a household, 28% of the respondents said only 2 persons lived in a household, 23% of the respondents said 3 people lived in a household, 22% of the respondents said 4 people lived in their households and 24% of the respondents gave different responses. The study also revealed the numbers of retired adults who lived in a household. The retired adults were about 19% of the sampled population. Therefore, this might translate to the fact that most of the population were working class. The distribution of people who were on seasonal income were shown in this study. About 32% of the respondents said it was one person in their households that was on seasonal income. The numbers of people who were employed on regular income were also evaluated. About 42% of the respondents said there was only 1 person on regular income in their households. The study evaluated the people's level of education. It was shown that about 46% of the sampled population had secondary education and 13% had high school. This is an indication that the communities would be receptive to education and broadcasting of environmental matters would thrive in those communities. Hence, the municipality is encouraged to devise means of reaching out to the community through various media such as, print media, campaign, and workshop. The household information is shown in Table 1.

The economic situation of a community can be best determined by the poverty level of the populace. Decision makers often used poverty rate to measure the economic conditions within communities and compare it with sections of the population. It is easier to determine the number of people who fall below poverty level through this means (Bishaw & Fontenot, 2014). The municipality must make it one of its goals to educate the public on environmental matters and to getting them involved in the implementation process of ZW plan and in the siting of recycling plant around them (Robinson & Nolan-Itu 2002).

From the study, it was observed that only 15% of the sampled population had businesses in the households while the remaining 85% did not have businesses in their households. From the results, 7% of the sampled population said they had grocery shops; 3% of the sampled population said they had sweet shops; 2% said they had hairdressing saloon; 3% said they were selling cooked food and about 85% gave no response.

From the data on the monthly income of the sampled polulation shown in Figure 2A, the average monthly income of 24 respondents was within ZAR 2000, for 11 respondents, it was within ZAR 5000 and for another 5 respondents, it was within the average of ZAR 10000. This showed that the living standard of the people within those communities was very low. A poverty line is when people are deprived of necessities of life because of the scanty resources at their disposal to maintain minimum standard of living. Some of the basic necessities of life include food, clothing, shelter, water, electricity, education and healthcare services. Most of the households earn less than $1/day, hence they could not afford the necessities of life. This is an indication that most of the people were living within the poverty line (Kamanou et al., 2005).

The average amount spent by households per month was shown in Figure 2B. From the 118 participants, 115 respondents responded and 3 people did not respond. Most times when it gets to do with finance, people tend to be very reserved. People who earned from ZAR 500 to ZAR 1700 formed about 50% (median) of the distribution. People who earned ZAR 1800 and above formed the remaining 50% of the entire distribution. People who earned from ZAR 7000 to ZAR 10,000 were less than 1% for each group. This showed that the level of poverty was high in those communities (USD 1 = ZAR 16.83 as of 24th July 2022). The mean and standard deviation of the distribution were calculated as 2587 and 1915 respectively.

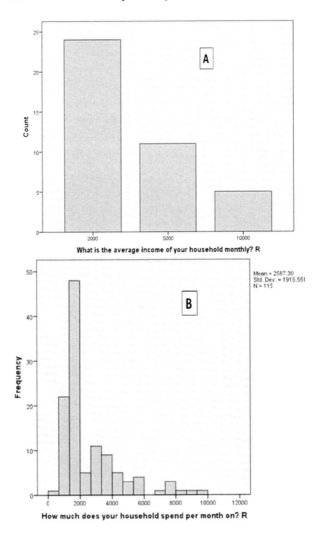

Figure 2. Average income of households per month (A) and amount spent by the household per month (B).

Major Concerns

High percentage of the respondents were concerned over the issue of air pollution from the landfill site around them and the poor waste collection services as shown in Figure 3.

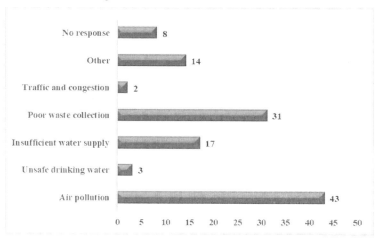

Figure 3. Environmental problem experienced by the residents.

Air pollution has become one of the major environmental problems being experienced by the populace since most of the waste disposed of is compacted during wet season, leading to emission (Lingan et al., 2014). Emission of chemicals, radiations and infectious diseases from the landfill poses risk to public health. Therefore, management of landfill site (LS) is to be strictly controlled in a way that its design and operation will not impact negatively on public health and the environment (Macklin et al., 2011). There is a need for the municipality to review its landfill management method and improve waste collection services especially in Naledi extension informal settlement where majority of the residents felt neglected because basic services were not offered to them by the municipality.

Household Waste Collection Service

For the waste collection services in the sampled communities, about 25% of the respondents said they had plastic bin inside their apartment, 10% said they had plastic bin at the entrance, 6% said they had plastic bin outside the apartment, 37% said they did not have refuse bin and 18% gave different feedback as shown in Table 2. The feedbacks given by the 18% are that they only stored their wastes in plastic bags, they did not have communal containers and others said they had refuse collection bin inside their apartments which were given to them by the municipality. With the bulk percentages of 37% of the respondents who were not having refuse collection bin, is an indication that illegal dumping of wastes might be inevitable in those communities.

Table 2. Public perceptions on household waste managment services in the sampled communities

Variables	Respondents					
Mode of storing household waste	Plastic bin inside	Plastic bin at entrance	Plastic bin outside	No refuse bin	Others	No response
	25%	10%	6%	37%	18%	4%
Condition of waste bins	Good condition		Damaged		No response	
	42%		6%		52%	
Duration in which bins are offered	1 year	2 years		Others		No response
	2%	13%		48%		37%
Mode of packaging waste for disposal	Only refuse bin	Refuse bin & plastic bags		Others		No response
	20%	20%		51%		10%
Days for refuse collection	Mondays		Thursdays		No response	
	63%		35%		2%	
Dumping sites within communities	Minor		Severe		Others	
	26%		9%		65%	
Rating of services rendered by service providers	Very satisfied	Satisfied		Not Satisfied		No response
	15%	31%		51%		3%
Reasons for lack of satisfaction	Infrequent collection	Unreliable collection	Location of containers	Number of plastic bags	Others	No response
	8%	27%	3%	7%	27%	29%
Destination of final disposal of collected wastes	Aware		Not aware		No response	
	26%		71%		3%	
Sufficient information on waste management	Yes		No		No response	
	9%		89%		2%	
Type of information needed by the communities	Waste collection schedule	Handling of different wastes	Where to channel complaints		Others	No response
	23%	12%	55%		6%	4%

Therefore, the Pikitup Johannesburg (Pty) Ltd (the municipality) has a lot of work to do by replacing the damaged refuse collection bins or providing new ones for those who were not having to prevent illegal dumping.

The study sought out the condition of waste bins. 42% of the respondents said their refuse collection bins were good, 6% said their refuse collection bins were damaged while 52% did not respond. With the 52% who did not respond, it is likely many of them might also be having damaged bins or not having any. It is therefore recommended that the Pikitup Johannesburg (Pty) Ltd (the municipality) speeds up action by deploying their workers to those communities to find out the need of the people and responding appropriately.

The study also found out the number of years in which the refuse collection bins had given to the people. From Table 2, it was observed that about 2% had received the refuse collection bin for just a year, 13% said it was for two years, 48% gave different responses other than the responses given earlier and 37% did not respond. With about 37% who did not respond, the municipality owe the communities a duty by finding to find out from them what their needs are so that provision can be made.

The study sought out how the people packaged their wastes for disposal. From Table 2, 20% said they often brought out only refuse collection bins provided by the municipality on refuse collection day, another 20% said they usually packed their wastes in refuse collection bins and plastic bags, 51% gave different responses in which majority of them said that only plastic bags were brought out on refuse collection day while 10% did not respond.

The study found out the days of the week in which waste collection services was taking place in those communities. For NEIS and Naledi RDP which formed 63% of the total participants, waste collection day was on every Monday and for Dobsonville RDP and Dobsonville Bond which formed 35% of the overall, the waste collection day was on every Thursday. The last 2% of the sampled population did not give any answer. So, with 98% of the respondents who were able to provide the days of their waste collection, is an indication that the municipality was offering services to the communities.

The study investigated any form of dumping site around and within the communities. As shown in Table 2, 65% of the sampled population said they only had minor dumping site around them, 9% said it was severe and 26% did not respond. For 65% who said they were having minor dumping sites around them, could possibly mean there were areas where illegal dumping is being carried out. This could be because they were not receiving any service or services offered were not enough. Hence, the municipality is requested to move into action by clearing every illegal dumping and positioning at least a community development worker (CDW) in the community whose duty is to be representing the municipality and educating the community on the need to stop illegal dumping and embrace SSWM.

Illegal dumping results when the public is not informed. It is the responsibility of Pikitup Johannesburg (Pty) Ltd to educate communities on environmental issues. Pikitup Johannesburg (Pty) Ltd could get the public and schools involved. The municipality should be able to educate the communities on the health hazard that could ensue from illegal dumping (Lotz-Sisitka et al., 2005).

The municipality could also formalise the activities of the informal sectors and scavengers who earn their living through the sales of waste items. Informal waste workers also known as reclaimers, scavengers or waste pickers are referred to as informal simply because they are not on contract, they do not have regular income, and no one recognises or cares about them making them very vulnerable. This informal waste sector contributes immensely to resource recovery and recycling in DCs. When the informal sectors are integrated into the waste management sector, the amount of recyclables would increase. The activity of informal sector makes good impact on the environment by providing a cleaner environment, reducing expenses that ordinary people are to incur because of WM services and these large groups of people are also making their income through this activity (Gerdes & Gunsilius, 2010). Municipalities are not to view waste pickers (reclaimers) as a problem but as a resource. When waste pickers are properly organized, they can become very active in the process of development. They can put an end to poverty at the base and they will begin to get good prices for their waste items especially when middlemen are eliminated in the process (Medina, 2008). Since households rarely recycle, reuse, or sell waste items, it is in the best interest of municipality to co-opt the informal sector into WM sector (Gunsilius, 2011).

The study also sought the opinion of the public on the level of satisfaction obtained with regards to the services offered by the municipality. As depicted in Table 2, 15% of the respondents were very satisfied with the services offered, 31% were satisfied and 51% were not satisfied. With about 51% who said they were not satisfied with the services offered to them; this might mean the municipality has not done enough in the discharge of quality services to the populace. Therefore, the municipality is encouraged to improve its services to the public.

From the results (Table 2), 8% of the sampled population said they were not satisfied because of infrequent collection, 27% said the reason for non-satisfaction was because of unreliable collection, 3% said it was because the location of the communal container was very far from them, 7% said the reason was because the number of plastic bags provided were not sufficient and 29% gave different responses. So, with about 27% of the respondents who said the issue they had was unreliable collection of their wastes, the municipality is encouraged to improve its services in those communities. Also, the 29% who did not respond might as well be having issues with their waste collection services. Hence, it is also recommended that the municipality deploys the CDWs to the communities for them to find out the issues that the people were experiencing and conveyed same to the municipality for prompt resolution.

Pikitup Johannesburg (Pty) Ltd is encouraged to strive for customer's satisfaction first and should make efforts in creating jobs for the unemployed youths in the community through green technology (Ethekwini, 2011). When the needs of the citizens are not met by the municipality especially in WM, it will be very difficult for the municipality to enjoy any form of cooperation from the citizens (Montalvo, 2009).

The study also found out from the communities whether they knew where their wastes were taken to for disposal. 26% of the respondents said they knew where their waste was taking to, for final disposal, 71% said they did not know and 3% did not respond. This shows that the knowledge of the people on environmental matters is insufficient.

The study established whether the people were given sufficient information on their waste collection services. 9% of the respondents said they had sufficient information, 89% of the respondents said they did not have any information and 2% did not respond. This also shows that the people of the communities were not well-informed. Information on who to reach in the event they had any issue with their waste collection services were also evaluated. 19% of the respondents said they knew who to contact should they had issues with their waste collection services, 77% said they did not know who to contact and 4% did not respond. With about 77% respondents who did not know who to contact should they be having issues with respect to their waste collection services is an indication that the people of the communities were not well-informed on matters relating to the environment.

On the kind of information that the communities would like to have. In Table 2, 23% of the respondents said they needed information on waste collection schedule, 55% said they needed information on where they could channel their complaints should they had any, 12% said they needed information on the proper way of handling different kinds of waste, 6% gave other information such as more plastic bags to be provided to aid source separation of wastes and 4% did not respond.

When municipalities create awareness and citizens are well-informed; they would support the services offered by municipalities. Hence, it becomes very paramount for municipalities to design better means of communication such as public campaign, media interventions etc. People are to be properly informed through campaigns including: radio jingles, media prints, pamphlets, adverts, banners, and posters (United Nations Conference on Trade and Development, 2014). One of the effective tools which municipalities can employ to get citizens involved and for improved service delivery is education campaign (Almarshad, 2015).

Willingness to Pay

The willingness of the residents to pay for their waste collection services was presented in the fifth section of the questionnaire.

From the sampled population, 41% of the respondents were aware that they are to pay for their waste collection services and 59% said they were not. Citizens are to be educated on the actual amount they were paying or which they are to pay for the services being offered to them since some of them thought because they were paying taxes, then they do not need to pay for the services.

Municipalities can also set up a contact centre which would serve as the organisational unit within the municipalities where citizens and businesses can channel their compliant and get answers to their queries and the existing contact centres can be improved upon since these would serve as an avenue to answer about 80% of queries forthwith. Municipalities are to be transformed into organisations that are more focused on customers. They should not hesitate to introduce changes in their work plans and service ideology (Millard, 2009).

The study further assessed the actual amount they the respondents were to be paying for their waste collection services. As shown in Figure 4, 20.3% of the respondents said it was supposed to be a free service since they were paying taxes, 0.8% said they were to pay ZAR 10, 14.4% said they were to pay ZAR 26, 1.7% said they were to pay ZAR 30, 0.8% said they were to ZAR 40, 0.8% said they were to pay ZAR 48, 5.9% said they were to pay ZAR 50, 1.7% said they were to pay ZAR 60, 3.4% said they were to pay ZAR 70, 0.8% said they were to pay ZAR 75, 4.2% said they were to pay ZAR 100, 3.4% said they were to pay ZAR 120, 0.8% said they were to pay ZAR 250, 0.8% said they were to pay ZAR 300 and 39.8% did not give any answer. The mean and standard deviation of the distribution were determined as 41.62 and 54.03 respectively. These are summary statistics measured for variables measured on a continuous scale. These tell us about the typical value in a data set and to what extent the other values vary around that typical value. Note: USD$ 1 = Rand (ZAR) 1 as of 27th July 2022.

The frequency of payment for waste collection charges by the respondents was considered. 19% said they were paying their waste charge regularly, 80% said they were not paying and 1% did not respond. It was observed that most of the respondents said they were not aware that they were to be paying and some said they did not need to pay since they were paying their taxes. This also shows that the people were not informed.

Municipalities are to continue to educate people and be creating awareness to the public as this could lead to change in peoples' attitudes towards the environment. Educating the public and creating awareness on the need to protect the environment would lead to sustainable development. "Sustainable development (SD) could simply be defined as harnessing natural resources to meet present needs without short-changing the succeeding generation to meet their own need" (Lephalale Municipality, 2011).

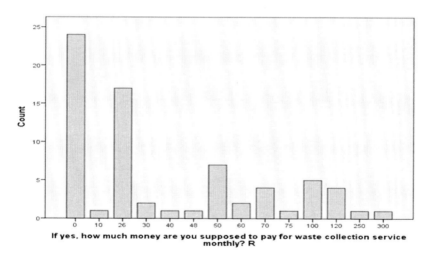

Figure 4. Amounts the residents think they should be paying.

The study sought to know whether the people would be willing to pay double the present amount for an improved service. 4% of the respondents said they would be willingly to pay more for improved services, 18% said they were not willing to pay more and 78% did not respond. This could be due to the level of poverty or lack of awareness of the people since it was only 4% of the entire sampled population that was willing to pay more for improved services.

Most of the households were not willing to pay more because they thought it might alter the budget they had already made for the running of their households` welfare (Eshun & Nyarko, 2011). As cities grow daily in DCs, so is waste generation but municipalities do not have the resources to provide quality services, although citizens are requesting for improved services (Hagos et al., 2012). The willingness to pay (WTP) more for a better service is linked to the level of civilization of the people and level of awareness created by the municipality (Addai & Danso-Abbeam, 2014).

Zero Waste (ZW)

ZW is a goal that is directed towards recovering of resources and safeguarding of the limited natural resources when waste is diverted from going to the incinerators and landfills. It involves minimization of waste, composting of waste, recycling, reusing and adjustment in the way people consume the limited natural resources and the likelihood for industries to redesign their products so that waste can be eradicated in the production processes (Allen et al., 2012). ZW as an alternative to the existing WM option was presented in the sixth section of the questionnaire.

From the sampled population, only 11% of the respondents indicated to have heard about ZW while the remaining 89% said they have not heard about it. This shows that the people need to be educated on separation of waste at source, reducing, reusing, recycling, and composting since all these formed parts of ZW.

The study sought the opinion of the people to know whether they would be willing to support ZW project. As depicted in Figure 5, 95% of the respondents were willing to support ZW, 4% were not willing and 1% did not respond. This shows that the implementation of ZW would thrive in those communities since the bulk of the respondents were willing to embrace the project.

How the communities will be supportive to the ZW project was also studied. 58% of the respondents said they would support ZW via separation of waste at source, 25% said it was going to be through recycling, 8% said it would be through reuse of waste items, 1% said it would be by burying the wastes and about 8% did not respond. This is also an indication that the people would embrace ZW since about 91% altogether were willing to support the project.

22% of the respondents affirmed that they were currently separating recyclables from other waste streams while the majority (78%) were not separating wastes into different streams. This shows that the municipality has a lot of works to do on sensitizing the people and encouraging them to begin to separate recyclables and most especially the organic wastes from other waste items since this will go a long way to extend the life of the landfill sites.

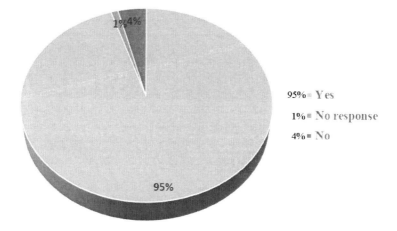

Figure 5. Willingness to support zero waste.

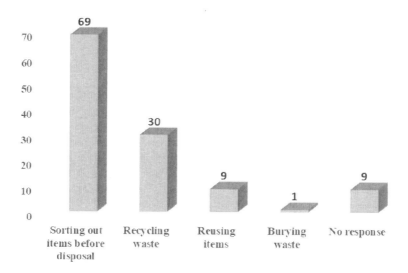

Figure 6. Ways to contribute to the success of zero waste.

Municipalities are to educate communities on the need to reduce amount of waste generated per person, to increase recycling rates, and to divert organic wastes from landfill facilities through source separation (Ayeleru et al., 2016). Organics are to be diverted from going to landfills and could be recovered via composting and anaerobic digestion. When organics are buried at landfill sites, they broke down and pose threats to public health and the environment since methane is usually released to the atmosphere and contributing greatly to global warming (The United States Composting Council, 2012). To recover resources from solid waste streams, separation at source should be of utmost priority. Source separation of waste has not been considered a priority in the developing nations. Waste separated at source can be recycled, composted, and can be converted to energy. Municipalities are having very great task and the task is to encourage members of communities to be actively involved in source separation of waste (Dagadu & Nunoo 2011).

The opinion of respondents towards their willingness to continue to separate recyclables showed that 24% of the respondents were willing to continue to separate recyclables, 3% were not willing while 73% did not give any answer. With the 3% who said, no and the other 73% who did not give any answer, showed the rate of recycling to be very low in those communities.

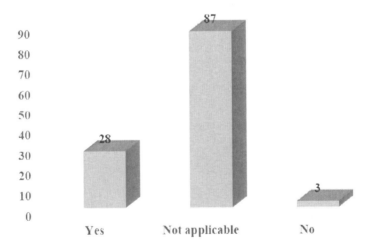

Figure 7. Willingness to continue to separate recyclables.

Where recycling of waste is very vibrant, there will be no need to extract or process new resources, hence fossil-fuel energy will be conserved and climatic impacts that are bound to occur when different method of WM are utilized will be avoided. Organic wastes are to be separated from inorganic wastes to prevent contamination of other recyclable items. Separation of organic waste at source is economical, more convenient, and very hygienic than disposing them (Evans, 2007). When an efficient source separation of organic waste is in place, resources will be saved. Food waste would be used as feeds in bio-digester which will later generate energy for heating or fuelling vehicles. In the long run, financial benefits will be obtained, and emission of carbon shall be reduced (Schmieder, 2012).

The study further sought the opinions of the people on the success of ZW. As shown in Figure 8, 93% of the sampled population agreed that ZW project could be successful, 4% of the respondents said it could not be successful and the remaining 3% did not give any answer. With the 93% of the respondents who indicated interest and were very optimistic that the project could succeed is a sign that the project would be viable.

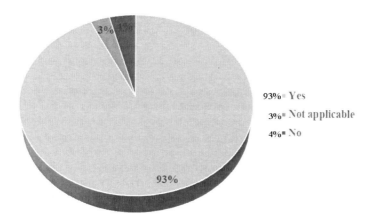

Figure 8. Opinion of the respondents on whether zero waste can succeed.

Conclusion

Utilizing engineering solutions alone to manage solid waste may not necessarily proffer the adequate solutions required to address the issue of poor solid waste management (SWM). In this study, the general perception of people towards environmental matter was evaluated using sociology of waste. It was discovered that the people were not conversant with the issues associated with their environment. Most of the people did not have the idea of where and how their generated wastes were being disposed of. Also, many of the respondents did not have access to basic waste management services such as provision of plastic bags or refuse bins for waste storage. Waste reduction, source separation and recycling of waste were at a minimal rate since the knowledge of the people was very poor. Some residents were not ready to pay for waste collection services and others were not ready to pay more for an improved service, but they were looking forward to an improvement in their waste collection services. However, most of the respondents were willing to support ZW project.

References

Addai, K. N., & Danso-Abbeam, G. (2014). Determinants of Willingness to Pay for Improved Solid Waste Management in Dunkwa-on-Offin, Ghana. *Journal of Agriculture and Environmental Sciences, 3*(1), 01-09.

Allen, C., Gokaldas, V., Larracas, A., Minot, L. A., Morin, M., Tangri, N., Tyler, B., & Walker, B. (2012). *On The Road to Zero Waste: Successes and Lessons from around the World.* Retrieved from Philippines: https://www.no-burn.org/wp-content/uploads/On-the-Road-to-Zero-Waste.pdf.

Almarshad, S. O. (2015). Municipal Awareness and Citizen Satisfaction: The Case of the Northern Borders in Saudi Arabia. *International Review of Management and Marketing, 5*(2), 94-101.

Ayeleru, O. O., Okonta, F. N., & Ntuli, F. (2016). *Characterization, management and utilization of landfill municipal solid waste: a case study of Soweto.* (Masters Dissertation). University of Johannesburg, Johannesburg.

Bishaw, A., & Fontenot, K. (2014). *Poverty: 2012 and 2013.* Retrieved from Washington, DC: https://www.census.gov/library/publications/2014/acs/acsbr13-01.html.

Bryman, A. (2004). *Social Research Methods*: Oxford University Press.

CIWM. (2014). *The Circular Economy: what does it mean for the waste and resource management sector?* Retrieved from Northampton: http://www.ciwm-journal.co.uk/downloads/CIWM_Circular_Economy_Report-FULL_FINAL_Oct_2014.pdf.

Coffey, M., & Coad, A. (2010). *Collection of municipal solid waste in developing countries*: UN-Habitat, United Nations Human Settlements Programme.

CoJ. (2006 – 2011). *The remaking of Soweto: End of Term Report* Retrieved from Johannesburg: http://www.joburg-archive.co.za/2008/pdfs/legacy_of_achievements/soweto_report.pdf.

Cooper, D. R., & Schindler, P. S. (2008). *Business Research Methods.* New York: McGraw-Hill.

CSIR. (2011). *Municipal waste management - good practices.* Retrieved from Pretoria: https://www.csir.co.za/sites/default/files/Documents/Waste_Management_Toolkit_0.pdf.

Dagadu, P. K., & Nunoo, F. (2011). Towards Municipal Solid Waste Source Separation at the Household Level in Accra. *Ghana. International Journal of Environment and Waste Management, 7*(3-4), 411-422.

DEA. (2011). *National Waste Management Strategy.* Retrieved from South Africa: https://www.environment.gov.za/sites/default/files/docs/nationalwaste_management _strategy.pdf.

Dulac, N. (2001). *The Organic Waste Flow in Integrated Sustainable Waste Management: Tools for Decision-makers: Experiences from the Urban Waste Expertise Programme.* Retrieved from Netherlands: https://www.ircwash.org/sites/default/files/Dulac-2001-Organic.pdf.

Eshun, T. B., & Nyarko, F. (2011). Willingness to Pay for Improved Waste Management Services: The Case of Tarkwa-Nsuaem. *Ghana Asian-African Journal of Economics and Econometrics, 11*(1), 187-196.

Ethekwini, M. (2011). *Ethekwini Municipality State of Local Innovation Report.* Retrieved from Durban: http://www.mile.org.za/Come_Learn/Knowledge_Management/Multi media%20Library/ABM%20Experiences%20Book/eThekwini%20Municipality-State%20of%20Innovation%20report.pdf.

Evans, T. (2007). *Environmental Impact Study of Food Waste Disposers.* Retrieved from England: https://www.insinkerator.jp/files/pdf/2007_england.pdf.

Gerdes, P., & Gunsilius, E. (2010). *The Waste Experts: Enabling Conditions for Informal Sector Integration in Solid Waste Management: Lessons Learned from Brazil, Egypt and India.* Retrieved from Deutschland: https://www.giz.de/en/downloads/gtz2010-waste-experts-conditions-is-integration.pdf.

Gunsilius, E. (2011). *Recovering Resources, Creating Opportunities: Integrating the Informal Sector into Solid Waste Management.* Retrieved from Germany: https://www.giz.de/en/downloads/giz2011-en-recycling-partnerships-informal-sector-final-report.pdf.

Hagos, D., Mekonnen, A., & Gebreegziabher, Z. (2012). *Households' Willingness to Pay for Improved Urban Waste Management in Mekelle City.* Retrieved from Ethiopia: https://www.jstor.org/stable/pdf/resrep14960.pdf?refreqid=excelsior%3A158b337b2 394df1a2bf2685ec294466e&ab_segments=&origin=&acceptTC=1.

Hisashi, O., & Kuala, L. M. (1997). *Sustainable Solid Waste Management in Developing Countries.* Paper presented at the 7th ISWA International Congress and Exhibition, Parallel Session.

Hox, J. J., & Boeije, H. R. (2005). Data Collection, Primary vs. Secondary. In *Encyclopedia of Social Management*: Elsevier.

Johnston, M. P. (2014). Secondary Data Analysis: A Method of which the Time Has Come. *Qualitative and Quantitative Methods in Libraries, 3,* 619 –626.

Kamanou, G., Morduch, J., Isidero, D., Gibson, J., Ivo, H., Ward, M., & Kakwani, N. (2005). *Handbook on Poverty Statistics: Concepts, Methods and Policy Use.* Retrieved from: https://unstats.un.org/unsd/methods/poverty/pdf/un_book%20final% 2030%20dec%2005.pdf.

Lephalale Municipality, L. (2011). *Integrated Waste Management Plan: Strategy: Waste Education and Awareness.* Retrieved from Botswana: http://www.lephalale.gov.

za/docs/reports/Lephalale_IWMP_Report_8_Education_and_Awareness_27_03_201 1_Version_71.pdf.

Lingan, B. A., Poyyamoli, G., & Boss, U. J. C. (2014). Assessment of Air Pollution and its Impacts near Municipal Solid Waste Dumping Site Kammiyampet, Cuddalore, India. *International Journal of Innovative Research in Science, Engineering and Technology, 3*(5), 12588-12593.

Lotz-Sisitka, H., Hamaamba, T., Kachilonda, D., Zondane, P., Kula, T., Olvitt, L., & Timmermans, I. (2005). *Makana Municipality Local Environmental Action Plan: Environmental Education and Training Strategy.* Retrieved from Grahamstown: https://www.ru.ac.za/media/rhodesuniversity/content/environment/documents/Enviro nmental_Education_and_Training_Strategy.pdf.

Macklin, Y., Kibble, A., & Pollitt, F. (2011). *Impact on Health of Emissions from Landfill Sites: Advice from the Health Protection Agency.* Retrieved from London: https://assets.publishing.service.gov.uk/government/uploads/system/uploads/attachm ent_data/file/334356/RCE-18_for_website_with_security.pdf.

Medina, M. (2008). *The informal recycling sector in developing countries: Organizing waste pickers to enhance their impact.* Retrieved from Washington D.C.: https://citeseerx.ist.psu.edu/viewdoc/download?doi=10.1.1.552.5069&rep=rep1&typ e=pdf.

Millard, J. (2009). Creating Customer Contact Centres: A guide for municipalities from Smart Cities. *European Journal of ePractice.*

Montalvo, D. (2009). Citizen Satisfaction with Municipal Services. *Americas Barometer Insights, 18*, 1-6.

Muller, M., & Hoffman, L. (2001). *Community Partnerships in Integrated Sustainable Waste Management.* Retrieved from Netherlands: https://www.ircwash.org/sites/ default/files/Muller-2001-Community.pdf.

Neuendorf, K. A. (2002). *The Content Analysis Guidebook.* United States of America: Sage Publications Inc.

Office of Planning and Institutional Assessment. (2006). *Innovation Insight: Using Surveys for Data Collection in Continuous Improvement.* Retrieved from United States: http://sites.psu.edu/opia/files/2018/05/No.-14-Using-Surveys-for-Data-Collection-in-Continuous-Improvement-qmwhwh.pdf.

Ogwueleka, T. (2009). Municipal solid waste characteristics and management in Nigeria. *Journal of Environmental Health Science & Engineering, 6*(3), 173-180.

Pokhrel, D., & Viraraghavan, T. (2005). Municipal solid waste management in Nepal: practices and challenges. *Waste Management, 25*(5), 555-562.

Robinson, L., & Nolan-Itu, W. (2002). *Pro-active Public Participation for Waste Management in Western Australia.* Retrieved from Western Australian: https://citeseerx.ist.psu.edu/viewdoc/download?doi=10.1.1.202.4768&rep=rep1&typ e=pdf.

Scheinberg, A. (2001). *Micro- and Small Enterprises in Integrated Sustainable Waste Management: Tools for Decision-makers: Experiences from the Urban Waste Expertise Programme.* Retrieved from Netherlands: https://www.waste.nl/wp-content/uploads/2019/04/tools_microent-1.pdf.

Schmieder, T. (2012). Food Waste at the University of Leeds–Maximising Opportunities. *Earth & E-nvironment, 7*, 201-231.

Simon, M. K., & Goes, J. (2013). *Scope, Limitations and Delimitations.* Retrieved from Seattle: https://www.academia.edu/6054602/SCOPE_LIMITATIONS_and_DELI-MITATIONS.

Siniscalco, M. T., & Auriat, N. (2015). *Quantitative research methods in educational planning.* Paris: UNESCO.

Snijkers, G., Haraldsen, G., Jones, J., & Willimack, D. K. (2013). *Designing and Conducting Business Surveys.* United States of America: John Wiley.

Squires, C. O. (2006). Public Participation in Solid Waste Management in Small Island Developing States. *Caribbean Development Bank, Barbados.*

The United States Composting Council. (2012). *Keeping Organics Out of Landfills.* Retrieved from Ronkonkoma: https://cdn.ymaws.com/www.compostingcouncil.org/resource/resmgr/documents/advocacy/advocacy_resources/keeping-organics-out-of-land.pdf.

The World Bank. (2009). *Developing a Circular Economy in China: Highlights and Recommendations.* Retrieved from Washington, D.C: https://openknowledge.worldbank.org/bitstream/handle/10986/18889/489170REPLACEM10BOX338934B01PUBLIC1.pdf?sequence=1&isAllowed=y.

United Nations Conference on Trade and Development. (2014). *Communication strategies of competition authorities as a tool for agency effectiveness: note/by the UNCTAD Secretariat.* (TD/B/C.I/CLP/28). UNCTAD secretariat Retrieved from https://unctad.org/system/files/official-document/ciclpd28_en.pdf.

Zhou, K., Bonet Fernandez, D., Wan, C., Denis, A., & Juillard, G. (2014). *A study on circular economy implementation in China.* Retrieved from: https://faculty-research.ipag.edu/wp-content/uploads/recherche/WP/IPAG_WP_2014_312.pdf.

Biographical Sketch

Olusola Olaitan Ayeleru

Affiliation: University of Johannesburg

Education: B.Eng., MTech., PhD Chemical Engineering

Business Address: Johannesburg, South Africa

Research and Professional Experience: Senior Researcher

Publications from the Last 3 Years:

1. Ayeleru, O. O.*, Fajimi, L. I., Oboirien, B. O., & Olubambi, P. A. (2021). Forecasting municipal solid waste quantity using artificial neural network and supported vector

machine techniques: A case study of Johannesburg, South Africa. *Journal of Cleaner Production*, 289, 125671.

2. Ayeleru, O. O. *, Sisanda Dlova, Ojo Jeremiah Akinribide, Freeman Ntuli, Williams Kehinde Kupolati, Paula Facal Marina, Anton Blencowe, Peter Apata Olubambi, 2020. *Challenges of plastic waste management in the sub-Saharan Africa: Waste Management, 110,* 24-42.

3. Ayeleru, O. O.*, Dlova, S., Akinribide, O. J., Olorundare, O. F., Akbarzadeh, R., Kempaiah, D. M., Hall, C., Ntuli, F., Kupolati, W. K. and Olubambi, P. A., 2020. *Nanoindentation studies and characterization of hybrid nanocomposites based on solvothermal process. Inorganic Chemistry Communications, 113,* p.107704.

4. Akinribide, O. J., Obadele, B. A., Akinwamide, S. O., Ayeleru, O. O.*, Eizadjou, M., Ringer, S. P., & Olubambi, P. A. (2020). Microstructural characterization and mechanical behaviours of TiN-graphite composites fabricated by spark plasma sintering. *International Journal of Refractory Metals and Hard Materials*, 105253.

5. Akinribide, O. J., Obadele, B. A., Ayeleru, O. O.*, Akinwamide, S. O., Nomoto, K., Eizadjou, M., Ringer S. P., and Olubambi, P. A., 2020. The Role of Graphite Addition on Spark Plasma Sintered Titanium Nitride. *Journal Materials Research and Technology.*

6. Akinribide, O. J., Obadele, B. A., Akinwamide, S. O., Bilal, H., Ajibola, O. O., Ayeleru, O. O*., Ringer, S. P. and Olubambi, P. A., 2019. Sintering of binderless TiN and TiCNbased cermet for toughness applications: Processing techniques and mechanical properties: A review. *Ceramics International, 45(17) Part A.*

Chapter 3

Evaluation of Domestic Solid Waste Management in the Qalqiliya District, Palestine

Issam A. Al-Khatib[1,*], PhD
Jafar A. A. Eid[2]
and Fathi M. Anayah[3], PhD
[1]Institute of Environmental and Water Studies, Birzeit University, Birzeit, Palestine
[2]Universal Institute of Applied and Health Research, Nablus, Palestine
[3]Faculty of Engineering and Technology, Palestine Technical University - Kadoorie, Tulkarm, Palestine

Abstract

This study describes the problems, issues, and challenges of municipal solid waste (MSW) management faced by 26 local authorities in Qalqiliya district of Palestine. Approaches of possible solutions that can be undertaken to improve MSW services are discussed. The study consists of a questionnaire for the public, and a survey with discussions with staff of local authorities involved in waste management. The study provides information on availability of MSW collection services and practices of waste disposal in Qalqiliya district. It was found that little or no consideration of environmental impacts was paid in the selection of MSW dumpsites which were not inspected or monitored consistently. Almost 46% of the local authorities disposed MSW in open random dumps without any treatment, and 15% of them disposed MSW in open random dumps and then burned it. All local authorities offered no pre- or post-employment training to the workers in MSW services, and hence they were usually exposed to serious threats. Small localities shared the

---

* Corresponding Author's Email: ikhatib@birzeit.edu.

In: Municipal Solid Waste Management and Improvement Strategies
Editor: Adam Fitz
ISBN: 979-8-88697-720-2
© 2023 Nova Science Publishers, Inc.

workers and vehicles of MSW collection. The number of available waste containers was little in most localities. The average frequency of MSW collection in several localities was 2.2 times per week. It was noticed that <9% of the total budget was allocated for MSW management, making the development of the sector challenging. Results also showed that 97% of the local residents were willing to pay more for better MSW services, 60% of them were willing to separate wastes into organic and inorganic voluntarily, and 19% of them were willing to separate MSW if funded by local authorities. Fortunately, 71.6% of the residents were ready to transform organic wastes into fertilizer products if they were adequately trained. Developing recycling societies does not only alleviate negative environmental impacts, it also promotes the sustainable management of MSW.

Introduction

General Background

Solid waste is defined as any material that has no value for people to store or use, and therefore, they prefer to dispose it of (Qusus, 1988; Kedir et al., 2019; Shanmugavel and George, 2021). Solid waste is classified into agricultural, industrial, pathological, hazardous, construction, and urban or municipal solid waste (MSW) (Shanmugavel and George, 2021). Solid waste can be segregated, transformed, reused, recycled, and recovered with a wide range of socio-economic and environmental advantages to the local community. Hence, solid waste management (SWM) becomes an important environmental health service that is essential to build healthy and sustainable cities and towns (Ahmed and Ali, 2004; Kedir et al., 2019).

In effect, SWM involves diverse options of MSW generation, segregation, collection, storage, transport, treatment, and final disposal (Al-Khatib et al., 2007, 2020; Thoni and Matar, 2019). Typically, people with higher incomes have higher standards of living, larger demands on goods and services, and as a result rising rates of waste generation. It is essential, therefore, to implement effective methods that reduce the volume and toxicity of generated waste, and utilize sound practices of SWM options which minimize MSW disposed of in landfills. SWM in developing countries such as Palestine is plagued by a number of problems, and potential solutions are constrained by financial, institutional, technical, and technological deficiencies (Al-Khatib et al., 2010, 2020; Ferronato and Torretta, 2019; Thoni and Matar, 2019). As a result,

SWM is highly dependent on donor funding (Mbuligwe et al., 2002), with the consequent non-sustainability of MSW service in Palestine (Thoni and Matar, 2019).

One of the most challenging SWM elements in developing countries has been identified as the final disposal of MSW (Talahmeh, 2005; Al-Khatib et al., 2007, 2010; Ferronato and Torretta, 2019). This mainly refers to the possible pollution of groundwater and surface water sources by leachate from unsanitary dumping sites (Al-Khatib et al., 2007, 2010; Kedir et al., 2019). Proper SWM practices at the source may not only minimize waste generation, it may also limit illegal dumping sites, prolong the use of sanitary landfill sites, and therefore, alleviate pollution associated with MSW (Mbuligwe, 2002; Tchobanoglous and Kreith, 2002; Talahmeh, 2005). The management of MSW at all stages of collection, transport, and disposal has been less than effective in most Palestinian localities (Al-Khatib et al., 2007, 2010).

Processing of MSW using suitable technologies must be a decision of the management system to make use of wastes and so to minimize burden on disposal options. The biodegradable wastes can be processed by composting, vermicomposting, anaerobic digestion or any other appropriate biological processing for stabilization (Tanaka, 1999; Pokhrel and Viraraghavan, 2005). Recyclable and recoverable materials in MSW should be utilized to conserve virgin resources, save energy, generate income, motivate sound attitudes and practices, and protect the environment (Tanaka, 1999; Al-Khatib et al., 2020; Shanmugavel and George, 2021). The Palestinian community lacks any segregation or treatment to the MSW, making more stress of the MSW system. In order to create successful solutions to such problems, there is a dire need for involving civil society institutions and raising public awareness (Mbuligwe et al., 2002; Kedir et al., 2019; Thoni and Matar, 2019).

The responsibility of SWM is divided among 13 joint service councils, 427 local government units (e.g., municipalities and village councils), and the United Nations Relief and Works Agency in 19 refugee camps in the West Bank of Palestine (Thoni and Matar, 2019). Collection and transportation of MSW in some Palestinian cities are relatively acceptable, but disposal is inadequate almost at all locations since the most common method of disposal is dumping and/or burning in open areas (Al-Khatib and Abu Safieh, 2003; Talahmeh, 2005; Al-Khatib et al., 2010; Thoni and Matar, 2019). The schedule and the method of MSW collection have to be clearly announced to the local community (Khan and Ahsan, 2003). The inadequate number and distribution of collection containers and the irregular collection schedule have encouraged the accumulation of solid waste in streets (Al-Khatib et al., 2007).

Modern machinery for collection and transportation of solid waste has been employed in most of the major municipalities, however, the number and maintenance of existing vehicles and trucks are not adequate to provide the service to all people and to empty containers as needed (Al-Khatib et al., 2007; Thoni and Matar, 2019). In many villages, the village council owns or rents a truck for the collection of solid waste, usually an agricultural tractor. Village councils rarely provide collection containers, and the village households store waste in plastic bags close to their houses or in the street until the collection truck passes by. A storage facility or bin shall be easy to approach and operate, not exposed to open atmosphere, aesthetically acceptable, and user-friendly (Huang et al., 2005). Given that manual handling of MSW has to be avoided, yet in urgent cases, it shall be carried out under proper precautions for the safety of workers (Milhem, 2004).

Landfilling shall be restricted to MSW that is neither reusable, recyclable, nor recoverable (Tchobanoglous and Kreith, 2002; Ferronato and Torretta, 2019). Dumping sites in the West Bank are not designed as sanitary landfills. These sites lack ground lining or leachate collection system to protect soil and groundwater. These sites are open and management is restricted to frequent burning of waste piles (Al-Khatib and Abu Safieh, 2003; Al-Khatib et al., 2007). As a rapid method of disposal, burning takes place in densely populated areas where clouds of smoke dominate (Al-Khatib et al., 2007; Thoni and Matar, 2019). Although there has been an improvement in collection procedures in some areas, the problem of waste disposal has not been solved, and therefore, solid waste disposal in unsuitable dumping sites has been creating environmental and human health problems (Al-Khatib and Abu Safieh, 2003; Talahmeh, 2005; Ferronato and Torretta, 2019).

During the last years, the quantities of MSW have dramatically increased. Currently heaps of wastes are common along the streets, alleys, and sidewalks. Despite the fact that residents have paid service fees and continually complained to city councils, very little has been done to provide storage bins and containers, or to collect the waste regularly and more frequently. As a result of these growing problems, this study aims to find out the factors contributing to the improper management of MSW. The four specific objectives addressed by this study are: 1. to assess people knowledge and attitudes regarding SWM, 2. to evaluate people knowledge about factors contributing to the improper SWM, 3. to introduce the subject of management of post-collection MSW, and 4. to determine the potential for MSW recycling and the factors which influence such decisions.

Study Area: Qalqiliya District

The name "Qalqiliya" goes back to Roman times, and European Mediaeval sources refer to it as "Kalkelie" used today by its contemporary residents. Qalqiliya district is located in the northern part of the West Bank with the green line as its western border situated about 12 km from the Mediterranean coast (Figure 1). Qalqiliya is a small district that is connected to the neighboring districts of Nablus from the east, Tulkarm from the north, and

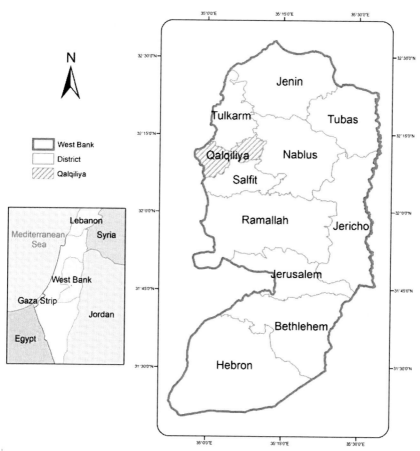

Figure 1. Qalqiliya district in the West Bank of Palestine.

Salfit from the south (Figure 1). In 2017, the total population of Qalqiliya district was 112,400 (PCBS, 2018), in a total area of 165.3 km², making a population density of 680 person/km² (MoLG, 2017). The number of employed persons in the public, civil society, and private sectors is 9,688 (PCBS, 2018). The total number of housing units is 28,635 with an average family size of 4.8 (PCBS, 2018). The district includes within its boundaries 5 municipalities, 30 villages (PCBS, 2018), and 13 Israeli settlements confiscating 36% of Qalqiliya land (ARIJ, 2007). Qalqiliya district is, as the rest of the West Bank, divided into Areas A, B, and C, which occupy 2, 25, and 73% of the total land in the district (ARIJ, 2007). Palestinians are assumed to have full sovereignty over land in Area A, Palestinians manage civil services while Israelis retain full responsibility for security in Area B, and Israelis retain full control over land in Area C (ARIJ, 2007). It is worth to mention that more than 90% of the population in Qalqiliya district live in Areas A and B, while Area C is dominated by agricultural lands (ARIJ, 2007). While agricultural activities generate only 22% of the total economy, the workers inside Israel represent the majority of labor force in the district (ARIJ, 2007).

Palestine enjoys typical Mediterranean climate conditions with hot, dry summers and cool, wet winters. The central highlands of 800 meters elevation have an average rainfall of 700 mm/year, while and the Dead Sea in the Jordan Valley which falls down 400 meters below sea level has less than 100 mm/year average rainfall (Anayah et al., 2021). The three major sources of freshwater in Palestine are the groundwater, springs, and harvested rainwater (Anayah et al., 2021). In the West Bank, the average value of relative humidity ranges from 50 to 70%, while average monthly temperatures range from 9 to 30°C (Anayah et al., 2021).

Methodology

This descriptive study was carried out in Qalqiliya district of the West Bank of Palestine. The population of the study consisted of two targeted groups. The first targeted group comprised of key persons in municipalities and village councils in Qalqiliya district. The second targeted group was all people residing in the district, and included all residents more than 15 years old in randomly selected study estates. A random sample of 5% of the households in each locality was taken.

Two semi-structured, yet simple, questionnaires were designed, pre-tested, and modified to collect required data. The first questionnaire was distributed to representative from each of the 26 municipalities and village councils (localities hereafter) in Qalqiliya district. The questionnaire targeted the key persons in the localities in order to get information about the present situation of MSW quantity, composition, generation, handling, treatment, disposal, and environmental impacts in the district. A total of 26 key persons were interviewed, one from each locality (i.e., municipality or village council).

The second questionnaire was for local households inorder to get the public opinion on the problems relating to SWM in Qalqiliya district. The questionnaire covered socio-demographic variables as well as variables relating to knowledge, attitudes, and practices on SWM among people living in the localities of Qalqiliya district. The interviews were conducted from door to door and questions were posed to either the family head or spouse (parents). In situations where none of them was present, either an adult son (over 18 years old) or a child (above 15 years old) was interviewed.

A total of 683 questionnaires were handed to the second targeted group of which only 667 were valid and can be processed for further analysis. The response rate of this questionnaire was high enough to exceed 97.6%. The questionnaires were distributed over Qalqiliya city (48%) and the villages of Qalqiliya district (52%) almost evenly. These percentages are close to the ratio of population size in Qalqiliya city (51683) to that in the district villages (60717) as indicated by the PCBS (2018).

Analysis of data was performed by the use of the Statistical Package for Social Science program. Descriptive statistics such as the average values and the ranges were computed. Appropriate tests of significance were performed to determine the relationship between socio-demographic variables and variables relating to knowledge, attitudes, and practices of local residents regarding SWM.

Results and Discussion

Perspectives of Locality Representatives

The representative from each of the 26 localities in Qalqiliya district were interviewed to better understand the system of SWM and performance of services provided to the public. Effective management of MSW requires public participation and full engagement from local communities. Local residents should feel the sense of responsibility and commitment, which is a key to the success of their communities (MoLG, 2011). Therefore, the first targeted group were asked about the commitment of the residents to pay fees collected for MSW services.

Figure 2. MSW collection workers without uniform or personal protective equipment in one of the localities of Qalqiliya districts.

The percentage of committed residents was >90% in 18 localities of Qalqiliya district, meaning that 31% of the localities had problems covering the basic costs of MSW collection. This was due to the absence of executive power of the localities as well as the economic decline residents had due to political instability. As a result, the quality of SWM was inappropriate and the citizens complained about uncollected refuse (or solid waste) along the streets.

Almost 14 out of the 26 localities could not easily find MSW workers, particularly when these workers had the chance to work for higher wages behind the Greenline (i.e., Israel) or within the settlements of the West Bank. The economic return is not the only reason behind this transition of the Palestinian workforce, social implications matter as MSW workers feel ashamed of the career and prefer to work in places where no one knows them. The health implications also matter as the risk of getting sick with communicable diseases is high for MSW workers while health insurance barely covers basic services.

Whilst occupational safety in the SWM system should be a top priority, strict precautions should be taken to minimize possible risks in the workplace. Personal protective equipment (PPE) has to be provided by local authorities for all MSW workers. The uniform is important for the safety of workers and might include a hat, special shoes, gloves, glasses, masks, and a plastic coat for extreme weather conditions. The representatives of the localities were asked whether MSW workers wear the uniform, and the answer of 15 of them was "No."

Figure 3. Scattering waste around a MSW container in Qalqiliya district.

It is unclear whether localities provided PPE or offered necessary training to the MSW workers. It is believed that the localities have financial and technical limitations regarding SWM and urgent interventions from competent authorities are needed. There is no question that no one cares about the safety of the MSW workers, however, it is sometimes about exposure risk assessment by employers and workers (Figure 2). Milhem (2004) investigated occupational health and safety hazards among MSW collectors in Bethlehem and Hebron districts. About 45% of MSW collectors in the two districts suffered from sore throat, cough, and high temperature, 28% of them suffered from diarrhea or bloody stool, 25% of them had shortness of breath, and 20% experienced skin diseases (Milhem, 2004).

Any mismanagement of the MSW system will lead to disastrous effects on public health and the environment. Out of the 23 representatives responded to this question, only 11 of them denied that residents burned MSW in the containers when it was accumulated for prolonged times. The residents burned the MSW in the containers not only to reduce its volume, it also solved the problems of ugly view and unpleasant odor released from accumulating waste. In addition to the damage resulting from burning of MSW containers, it also affects the health and safety of residents in the surrounding area by polluting the air and causing many respiratory problems such as shortness of breath.

Another health and environmental problem was the scattering MSW around containers especially full ones. Only 3 out of 22 representatives responded to this question denied the presence of such a problem in their localities. The three localities emptied MSW containers every day and they had sufficient number of containers spread all over the locality to correspond to collection frequency.

Unfortunately, this was not the case in the remaining 19 localities surveyed here. This was due to the low collection frequency in these localities and the long distance between MSW containers which were insufficient at the first place. Representatives added that they sometimes found waste around MSW containers when there was a delay in the collection time. Other reasons included the disposal practices of some residents who left waste sacks around the containers or due to wind that blew and scattered the waste from uncovered containers as shown in Figure 3.

Table 1 shows multiple features of the existing SWM system in the 26 localities of Qalqiliya district. These features include the percentage of households covered by MSW collection services, the MSW collection frequency, the monthly salary of MSW collection workers, and the number of SWM workers. The results revealed that the percentage of households covered by MSW collection services ranged between 80 and 100% with an average value of 96.9% compared to 90.7% in the year 2005 (PCBS, 2005). The non-coverage of the remaining households was mainly due to their locations and the absence of paved roads to reach these households.

A wide variation was noted in MSW collection frequency, which ranged from 1 to 6 times per week with a low average value of 2.2 times per week. This low average frequency of MSW collection needs immediate interventions from SWM system. Among all localities in the district, Qalqiliya city had the highest MSW collection frequency of 6 times/week. This was attributed to the broader scale of services offered by the cities, including street cleaning. This was also the reason behind the large number of SWM workers in Qalqiliya city as depicted in Table 1.

Analyzing the questionnaire of the local authority revealed that 65.4% of the localities collected less than 2 tons of waste per day. This led us to the fact that the population density in the localities of Qalqiliya district was really low. The low collection rate was primarily due to resource constraints including the lack of MSW collection vehicles, and the financial and physical limitations.

The study proved that 88.5% of the localities shared MSW collection vehicles and/or collection crews under a SWM system known as the "joint service councils." Sometimes, seven localities shared the same vehicle and the same crew which alleviated some of the financial burden borne by the smaller villages to provide MSW services to their residents. This explained the substantial increase in the coverage rate of MSW collection services in the past few years.

Table 1. Features of SWM system in the localities of Qalqiliya district

No.	Locality	Percentage of households covered by MSW collection services	MSW collection frequency	Salary of MSW collection workers	Number of SWM workers	
		(%)	(time/week)	(NIS/month)	From	To
1	Ad-Dab'a	100	1	1700	2	7
2	Al-Funduq	90	2	1750	2	3
3	An-Nabi Elyas	100	3	1000	3	4
4	Azzoun	100	3	1150	7	7
5	Azzoun Atma	100	2	1700	2	7
6	Baqa Al-Hatab	99	2	1200	3	6
7	Beit Amin	100	2	1500	3	3
8	Far'ata	95	1	1200	3	6
9	Habla	95	3	1200	3	3
10	Hajja	100	2	1200	3	6
11	Immatin	95	2	1200	3	6
12	Isla	100	1	1000	3	4
13	Izbat Al-Ashqar	100	2	1700	2	7
14	Izbat At-Tabib	90	1	1700	2	7
15	Izbat Jal'ud	98	1	1700	2	7
16	Izbat Salman	99	2	1700	2	7
17	Jayyous	90	3	1000	3	4
18	Jinsafut	80	2	1750	2	3
19	Jit	100	2	1200	3	6
20	Kafr Laqif	95	2	1200	3	6
21	Kafr Qaddum	98	3	1750	2	3
22	Kafr Thulth	100	3	1500	3	3
23	Qalqiliya	95	6	1350	80	80
24	Ras Atiya	100	2	1700	2	7
25	Sanniriya	100	2	1500	3	3
26	Sir	100	2	1000	3	4
Minimum		80	1	1000		
Average		96.9	2.2	1406		
Maximum		100	6	1750		
Sum					149	209

It was noticed that most of the MSW collection vehicles were 3-ton compactors (Figure 4) that took one delivery from each locality every two or three days. Sometimes the compactor was filled with MSW before emptying all waste containers in the same locality, so these containers will not be emptied till the next time. This spread bad odors all the weekdays and would be a suitable environment for insect breeding, not to mention the ugly view. Consideration of the composition of MSW can help make the correct choices in importing MSW handling equipment. For example, there is no need to import compactor trucks, which are suitable to less dense MSW; dense MSW, which needs no compaction but just needs hauling, trucks that might be even cheaper.

Figure 4. Three-ton MSW compactor heavily used in Qalqiliya district.

(a)   (b)

Figure 5. MSW dumpsites in the area of: (a) Qalqiliya city, and (b) Jayyous village.

Five MSW disposal sites of different sizes and disposal rates were visited. Some of these dumpsites were located at environmentally sensitive areas. Other dumpsites were in the middle of agricultural lands and on top of groundwater aquifers (Figure 5). Qalqiliya is rich with groundwater resources and it is believed that a few of its dumpsites are in the middle of a water-bearing layer.

Out of the 26 localities of Qalqiliya district, 7 of them basically burned MSW while 3 of them had semi-covered dumping sites. Almost 61.6% of the localities disposed of MSW in random open dumpsites of which one-fourth was burned lacking proper health and safety requirements. The reason why localities used the burning method was volume reduction, in addition to the limited budget that did not allow for any further treatment techniques or sound disposal options.

As a result, localities may perform their MSW services according to the resources available from collection fees leading to deterioration of MSW services. Because the localities had limited financial resources, uncontrolled dumping may be the only option available now because it was the cheapest type of land disposal. However, it should be noted that such practices put public health and the environment at risk from underground and surface water contamination, toxic smoke and waste blown by wind, vectors, etc.

SWM should put an end to these practices which are environmentally unacceptable and cut off the useful life of a disposal site. Once the disposal site is filled out, new ones must be found mostly at a longer distance and higher transportation cost. Still, an effective strategy to make disposal sites have the longest possible life should focus not only on technical operations at the site, but also on waste diversion. MSW diversion includes source reduction, recycling, recovery, and waste transformation through composting.

Moreover, it was noticed that the dumpsites were not protected from the entrance of children or animals which was a serious problem because children searching the waste for valuable metals and materials from discarded items, while animals eat polluted waste from the dumpsite. The medical waste generated in Qalqiliya district, is a serious threat to the population of the area. The results showed that 45.8% of the localities disposed their medical waste with MSW without any treatment, and this puts the children and animals under the risk of being infected with communicable diseases.

A little percentage of generated medical waste was properly treated before final disposal. Only one locality burned the mixed waste, while 5 localities collected medical waste and burned them in a special incinerator. Most medical waste was mixed up with MSW and disposed of together in the same containers. This wrong practice exposed MSW collection workers and scavengers to a great danger of disease infection.

As for the ownership of the dumpsite land, 5 localities owned it, 20 of them rented it, and one of them only used a public land. Renting the land of the dumpsite increased the financial load over localities, participated in decreasing MSW services offered to residents, and escalated environmental problems. Every locality wanted a closer dumpsite location to reduce transportation costs. This practice caused the spread of dumpsites all over the district without any treatment and this resulted in major environmental problems. Allocating sufficient financial and technical resources is an essential key to the success of the SWM system in Palestine.

As shown in Table 2, the percentage of total budget allocated for SWM was less than 3% in 11 localities of Qalqiliya district. Compared to the upper threshold of 9% of the total budget for SWM, only 2 out of the 26 localities had a higher percentage. The budget for SWM was between 2 and 8% of the total budget for municipalities in seven districts (including Qalqiliya) in the north and middle of the West Bank (Al-Khatib et al., 2007). It is very small percentage of the total budget when compared with SWM budget in a developing country such as Sri Lanka in which SWM budget ranged between 12 and 20% of the total budget (DCS, 1998).

The SWM fees were collected on a monthly basis with electricity bells in 92% of the localities, or yearly with a special SWM bell in the other localities. When representatives were asked about the fees that localities collected from residents, 85% of them said that fees fairly covered SWM expenses, while 15% of them said that fees were not sufficient. The majority of fees collected were spent on salaries to staff as well as fuel and maintenance for MSW vehicles (Al-Khatib et al., 2010; Thoni and Matar, 2019). This small SWM budget resulted in lower level of services, less collection frequency, no street litter collectors, and absence of suitable methods for final waste disposal. Since localities had limited financial resources, uncontrolled dumping may be the only option available at the moment because it was the cheapest type of waste disposal.

Table 2. Percentage of total budget allocated for SWM in Qalqiliya district

SWM budget/Total budget (%)	<3	3 to 6	6 to 9	>9
No. of localities	11	9	4	2

Socio-Demographic Characteristics of Household Respondents

Table 3 shows the socio-demographic characteristics of the targeted respondents to the questionnaire. Approximately 88.3% of the families had an income of less than 3000 New Israeli Shekel (NIS), given that 1.0 US$ = 4.32 NIS at the time of the study. More than 96% of the respondents were adults above 18 years old, either family heads or large sons. The family size was ≤8 persons, which is higher than national average of the PCBS (2018), for around 77% of the respondents. Almost 88% of the respondents lived in households of ≤4 rooms for >10 years.

Table 3. Socio-demographic characteristics of the targeted respondents

Factor	Category	Frequency	Percentage (%)	Sum
1. Family income (NIS/month)	<1000	208	31.5	660
	1000 – 2000	271	41.1	
	2000 – 3000	104	15.8	
	3000 – 4000	46	7.0	
	>4000	31	4.7	
2. Type of respondents	Family head	530	79.5	667
	Adult son	112	16.8	
	Child	24	3.6	
	Others	1	0.1	
3. Family size	<5	176	25.8	681
	5 – 8	346	50.8	
	9 – 12	136	20.0	
	>12	23	3.4	
4. No. of rooms	<3	171	25.1	682
	3 – 4	427	62.6	
	>4	84	12.3	
5. Duration of residence in localities (years)	<10	72	11.2	645
	10 – 25	172	26.7	
	26 – 35	108	16.7	
	>35	293	45.4	
6. Education level	Illiterate	17	2.5	681
	Elementary	70	10.3	
	Preparatory	165	24.2	
	Secondary	257	37.7	
	Post-secondary	172	25.3	

Most people had lived at the same house for a long time because they wanted to stay together as a family, and this increased the population density per household. A few families had changed the place where they lived due to marriage or work requirements. It was obvious that the population density was high, while income was low for most families in Qalqiliya district. This would negatively affect the attitudes and practices of the local residents towards MSW. Good examples in fact included increasing littering and indiscriminate disposal of MSW in the local communities. As depicted in Table 3, although the sample represented all levels of education, but 63% of the respondents had a secondary (high school) or higher education level. This indicates the high likelihood that locals are willing to positively change their behaviors towards public health and the environment.

Perspectives of Household Respondents

The perspectives of the local residents were revealed through the analysis of the 667 questionnaires collected from Qalqiliya district. It is important to understand the attitudes and practices of the local residents and explain their perspectives towards SWM. One of the questions posed to the respondents was about the family member who is in charge of waste disposal. Regarding the family member who disposed waste into MSW containers, 17.1% of the respondents admitted that children usually threw waste near MSW containers. While children should not be in charge of waste disposal, they still practiced this and were blamed not to put waste into MSW containers.

Hence, waste accumulating near MSW containers resulted in unsightly views, bad odors, and animals (especially cats, dogs, rodents, and insects) searching for food waste, scattering waste in all directions, and spreading communicable diseases (Figure 6). MSW containers should be tall enough to prevent children from exposure to dangers, and therefore, adults are responsible for waste disposal into MSW containers. In addition, some of these containers come with a lid, making it difficult for the child to open and drop the waste bag into the container at the same time.

This was why some residents were afraid to put MSW containers in front of or close to their homes. In figures, almost 183 (27%) of the respondents were disturbed from putting a waste container close to their homes. When the 183 respondents were asked about the main reason behind such an attitude, 59% of them complained from the stinky odors, 13% of them were worried from waste accumulation, and 9% were concerned about unsightly views and insects combined with bad odors.

Respondents were asked about their willingness to pay (WTP) to receive a better MSW service in their locality. It is worth to mention that the educational level of respondents was found to be statistically significant to their WTP for better MSW services (p-value = 0.001). Only 18 respondents were against any additional fees for a better MSW service (Figure 7). The results showed that 97.3% of the respondents were willing to pay more for better MSW services in their localities.

Figure 6. Cats scattering waste and searching for food in a MSW container.

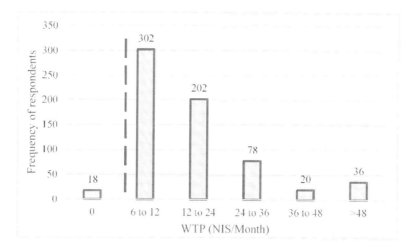

Figure 7. The frequency of respondent's willingness to pay for better MSW services.

From Figure 7, it is clear that the WTP for 46% of the respondents approached 12 NIS/Month, while it reached 24 NIS/Month for 31% of them. These findings have to be carefully read and understood by competent authorities in the SWM system. The WTP is an important measure that explains the existing situation of insignificant SWM in the localities of Qalqiliya district. Policy and decision makers have to take these findings into account when it comes to any development in the SWM system.

When respondents were asked whether the distance to the nearest MSW container was suitable or not, 70.7% of the respondents were satisfied about the walking distance to the nearest MSW container. About 33.1% of these respondents preferred that MSW container was far away from their homes because of the unsightly views, bad odors, and spread of insects and rodents. The remaining 29.3% of the respondents complained from either the distance to (17.2%) or the absence of (12.1%) a MSW container in their neighborhoods. Figure 8 indicates that 52.4 and 31.9% of the respondents preferred the nearest MSW container to be within 20-meter and 50-meter distances from their homes, respectively. It is uncommon in domestic culture to dispose of waste by vehicle, particularly when a child performs this task as mentioned earlier. People wanted the MSW container to be far away within the distance needed to prevent the negative effects.

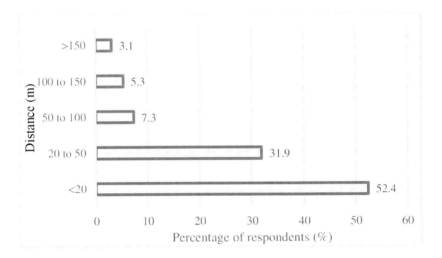

Figure 8. Acceptable walking distance to reach the nearest MSW container.

At the end, it is important to understand how satisfied the residents were about the existing MSW services. The results showed that 38.6% of the respondents were fully satisfied with existing MSW services, 44.7% of them were relatively satisfied, and 16.7% of them were unsatisfied at all. This little portion of respondents were unsatisfied with existing MSW services because of the irregular collection of waste, the deficiency or absence of MSW containers, and the poor maintenance of these containers. It is worth to mention that the family income (p-value = 0.022) and size (p-value = 0.001) of respondents were found to be statistically significant to their satisfaction with the existing MSW services.

Initiatives to Better Manage Municipal Solid Waste

Future initiatives for a successful SWM system in Qalqiliya district are to be explored in this section so that the willingness of residents to participate in pro-environmental programs and activities is investigated. Proper planning is essential for an effective management to the MSW sector. The first initiative was about sorting waste into five categories: organic, solid, liquid, recyclable, and hazardous waste. If the first initiative did not work for the local community, the second initiative proposed would be to sort waste into two categories only: organic versus inorganic waste. It is worth to mention that the educational level of respondents was found to be statistically significant to their willingness to separate waste into five categories (p-value = 0.006).

As shown in Figure 9(a), 42% of the respondents agreed to separate waste into five categories to achieve a sustainable SWM system, 24% of them agreed if there was some seed money to establish the project and encourage a larger participation of public, and 34% of them disagreed to separate waste into five categories. When those disagreed to participate in the initiative were asked about the reason behind their decision, they believed it is a useless process that requires enough time and space. The authors believe that the spark to ignite such initiatives is to allocate sufficient financial and technical support, alongside with raising awareness and educating public about the positive impacts of such pro-environmental programs.

As shown in Figure 9(b), the respondents were asked about their willingness to separate MSW into two categories: organic and inorganic waste. The percentages of respondents agreed to participate in this initiative increased which was expected because of the less resources and efforts needed for two types of waste compared to five. Fortunately, 60% of the respondents agreed to voluntarily separate waste into two categories, while 19% of them accepted to participate if there was some seed money.

The percentage of respondents disagreed to separate waste into five categories decreased by around 50% when they were asked to sort waste into two categories only. When respondents disagreed to participate in these initiatives were asked about the reasons behind their decision, some of them admitted that they did not recognize the difference between organic and inorganic waste, while others were afraid from associated risks of the sorting process. It is believed that such initiatives need to be introduced to the community step by step to assure larger public participation and commitment which will lead to the success of the program.

In the domestic culture of Qalqiliya district, people almost always cock and eat at home, generating larger amounts of organic (or food) waste. This is why several studies discussed options Palestinians do have to better manage their food resources and wastes (Bencivenni, 2017, Thoni and Matar, 2019). In a neighboring district, namely Nablus, organic waste was found to be 65.1% of the overall composition of MSW (Al-Khatib et al., 2010). Even in some workplaces such as healthcare centers of Jenin district, food waste represented 25% of the healthcare waste composition (Al-Khatib et al., 2020). Hence, a better management is needed to convert this type of waste into a resource.

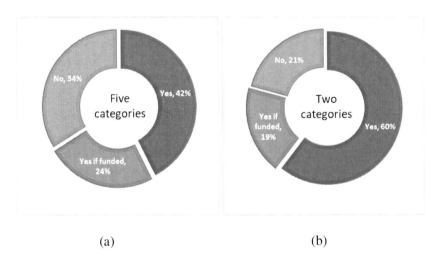

(a)   (b)

Figure 9. Willingness of respondents to separate MSW into: (a) five categories, and (b) two categories.

The residents were asked about how they disposed of their food waste. The result revealed that 57% of the respondents got rid of their food waste with other types of waste in MSW containers, and 31% of them used food residues to feed their animals. Raising animals need larger tracts of land which exist in the countryside of Qalqiliya district. Surprisingly, a few respondents recycled their food waste into compost by direct mixing of the waste with soil to produce an organic fertilizer, yet the percentage of them did not exceed 2%.

Composting does not only protect the environment, it also helps residents improve soil fertility and crop productivity in their gardens. Local authorities should explore possibilities to motivate local people composting domestic organic waste. If implemented properly, home composting would be a sustainable solution for small local authorities with larger rates of food waste generation. Although the production rate of domestic organic waste is low, it is to some extent constant throughout the year making such investment sustainable in the long-term.

In effect, the respondents' willingness to transform organic waste into natural fertilizer was examined. The results showed that 71.6% of the respondents had the willingness to transform organic waste into natural fertilizer if they were trained by specialists. As for the remaining 28.4% of the respondents who rejected this idea referred that to several reasons; a few of them had neither backyards nor time, while others used different types of fertilizers. It is worth to mention that the educational level and family income of respondents were found to be statistically significant to their willingness to transform organic waste into natural fertilizer (both have p-value = 0.004).

Conclusion

Localities (i.e., municipalities and local councils) employed workers in municipal solid waste (MSW) services without proper training, and therefore, they were always exposed to serious health and environmental threats. Localities often experienced financial difficulties in meeting the large payment of wages, fuel and maintenance of MSW vehicles, etc. Most local authorities had become economically constrained in offering efficient management of MSW. The rapid population growth had overstretched the capacity of local authoritiesto adequately provide quality MSW services. In spite of the high coverage rate of waste services to the residents of Qalqiliya district, the average frequency of MSW collection in the 26 localities was 2.2 times per week which is low enough.

The building of MSW collection and disposal capacities needs a broader approach to address the improvement of local infrastructure; including the need to upgrade roads leading to dumpsites. Many localities were blamed for breakdowns of their MSW collection vehicles and the accumulation of waste in and near MSW containers especially in winter and crowded areas. Localities, though poor, should develop area-specific solutions to their solid waste management (SWM) problems.

Community involvement through neighborhood groups of people from middle and higher income groups and businessmen can provide the needed solution in mobilization of community-based efforts. Clean neighborhood groups can mobilize financial resources and engage private groups or hire private trucks to occasionally collect and dispose MSW from their neighborhoods. Other measures include cultivation of a sense of clean environment through clean community awareness programs. These can go a long way in sensitizing people to keep the surrounding environment clean.

Regular activities and campaigns such as cleanup of neighborhoods, schools, parks, and roadsides can be effective in changing the residents' attitudes even among poor communities. In general, the proper MSW is determined by the attitudes of people towards waste, such as the ability to refrain from littering and indiscriminate disposal of MSW. Socio-economic characteristics may determine attitudes such as the ability/willingness to recycle MSW. These attitudes, however, may be positively influenced by awareness-building campaigns and educational measures. In a word, it is the desire of local people to keep their locality clean.

In the 26 localities studied, it was found that little or no consideration of environmental impacts was paid in the selection of dumpsites, including those currently in use. Convenience or availability of land took priority in the selection to the location of a dumpsite. Inspection and monitoring of dumpsites was not consistent, 46.2% of the localities disposed MSW in open random dumps without any further treatment, while 15.4% of them disposed of MSW in open random dumps and then burned it.

No sanitary practices such as application of daily soil cover or fencing were practiced in any of the dumpsites studied. One exception was in Qalqiliya dumpsite, the municipality covered the waste with a thin layer of soil on a daily basis. None of the dumpsites met the basic requirements in protecting groundwater from pollution as there were no liners to prevent leachate penetration. Overall, public-civil-private partnership is crucial for a better planning and management of MSW system in Palestine.

References

Ahmed, S. A., & Ali, M. (2004). Partnership for solid waste management in developing countries: Linking theories to realities, *Habitat International, 28*, 467–479.

Al-Khatib, I. A., Arafat, H. A., Basheer, T., Shawahneh, H., Salahat, A., Eid, J., & Ali, W. (2007). Trends and problems of solid waste management in developing countries: a case study in seven Palestinian districts. *Waste Management, 27*(12), 1910–1919.

Al-Khatib, I. A., Khalaf, A., Al-Sari, M. I., & Anayah, F. (2020). Medical waste management at three hospitals in Jenin district, Palestine. *Environmental Monitoring and Assessment, 192*(10), 1–15.

Al-Khatib, I. A., Monou, M., Abu Zahra, A. S., Shaheen, H. Q., & Kassinos, D. (2010). Solid waste characterization, quantification and management practices in developing countries. A case study: Nablus district - Palestine. *Journal of Environmental Management, 91*(5), 1131–1138.

Al-Khatib, I., & Abu Safieh, R. (2003). *Solid waste management in emergency: A case study from Ramallah and Al-Bireh municipalities.* Institute of Community and Public Health, Birzeit University, Palestine.

Anayah, F., Al-Khatib, I. A., & Hejaz, B. (2021). Assessment of water and sanitation systems at Palestinian healthcare facilities: Pre and post COVID-19. *Environmental Monitoring and Assessment, 193*(41), 1–22.

ARIJ, (2007). *Geopolitical status in Qalqiliya Governorate.* Applied Research Institute - Jerusalem, Bethlehem, Palestine.

Bencivenni, S. (2017). *Food loss and waste in Palestine: A pilot study of the FAO methodology.* Master Thesis. Polytechnic University of Milan, Italy.

DCS, (1998). *Municipal solid waste statistics.* Department of Census and Statistics, Colombo, Sri Lanka.

Ferronato, N., & Torretta, V. (2019). Waste mismanagement in developing countries: A review of global issues. *International Journal of Environmental Research and Public Health, 16*(6), 1060.

Huang, G. H., Linton, J. D., Yeomans, J. S., & Yoogalingam, R. (2005). Policy planning under uncertainty: Efficient starting populations for simulation-optimization methods applied to municipal solid waste management, *Journal of Environmental Management, 77*, 22–34.

Kedir, A. K., Eliku, T., & Gebre, G. D. (2019). Solid waste management practices and challenges: The case of Halaba Town in Southern Ethiopia. *Journal of Environment and Earth Science, 9*(9), 14–24.

Khan, I. H., & Ahsan, N. (2003). *Textbook of solid waste management.* CBS Publisher and Distributors, New Delhi, India.

Mbuligwe, S. E. (2002). Institutional solid waste management practicesin developing countries: A case study of three academic institutions in Tanzania. *Resources, Conservation and Recycling, 35,* 131–146.

Mbuligwe, S. E., Kassenga, G. R., Kaseva, M. E., & Chaggu, E. J. (2002). Potential and constraints of composting domestic solid waste in developing countries: Findings from a pilot study in Dar Es Salaam, Tanzania. *Resources Conservation and Recycling, 36,* 45–59.

Milhem, A. K. (2004). *Investigation of occupational health and safety hazards among domestic waste collectors in Bethlehem andHebron districts.* Master Thesis, An-Najah National University, Nablus, Palestine.

MoLG, (2011). *Promoting and institutionalizing public participation in local government units' affairs.* Ministry of Local Government, Ramallah, Palestine.

MoLG, (2017). *Geographical information management system in Palestine (GeoMOLG).* Ministry of Local Government, Ramallah, Palestine.

PCBS, (2005). *Household environmental survey: Main results.* Palestinian Central Bureau of Statistics, Ramallah, Palestine.

PCBS, (2018). *Preliminary results of the general census of population, housing and establishments 2017.* Palestinian Central Bureau of Statistics, Ramallah, Palestine.

Pokhrel, D., & Viraraghavan, T. (2005). Municipal solid waste management in Nepal: Practices and challenges. *Waste Management, 25*, 555–562.

Qusus, S. K. (1988). *Composition and generation rate of the solid waste of hospitals and medical laboratories in Amman.* Master Thesis, Jordan University, Amman, Jordan.

Shanmugavel, G., & George, B. (2021). *Textbook of public health and hygiene for undergraduate and postgraduate students.* Darshan Publishers, Tamil Nadu, India.

Talahmeh, I. (2005). *Good planning for sanitary landfill: Hebron district as a case study.* Master Thesis, Birzeit University, Birzeit, Palestine.

Tanaka, M. (1999). Recent trends in recycling activities and waste management in Japan. *Journal of Material Cycles and Waste Management, 1*, 10–16.

Tchobanoglous, G., & Kreith, F. (2002). *Handbook of solid waste management.* McGraw-Hill, New York, USA.

Thoni, V., & Matar, S. (2019). *Solid waste management in the occupied Palestinian territory: West Bank including East Jerusalem and Gaza Strip.* CESVI Overview Report, Bergamo, Italy.

Chapter 4

Food Waste Issues, Potential Applications, and Recovery Plans in South Africa: A Review

Olusola Olaitan Ayeleru[1,2,*], PhD
and Peter Apata Olubambi[1], PhD

[1]Centre for Nanoengineering and Tribocorrosion (CNT), University of Johannesburg, Johannesburg, South Africa
[2]Conserve Africa Initiative, Osogbo, Osun State, Nigeria

Abstract

Food waste generation has continued to be on the increase globally, notably in low-income countries. The estimated global population increases of about 200,000 people daily translates to approximately 100 persons daily for Africa, thus expanding food waste generation. The food waste generated is estimated to be around one-third (approximately ~1.3 billion tonnes) of the total quantity of food produced globally and its economic analysis is valued at ~USD\$ 800 billion. Moreover, the amount of food waste generated in sub-Sahara Africa is about 40% of the total tonnes of food produced (~100 million tonnes) and ~10 million tonnes in South Africa yearly. The continuous increase in the amount of food waste generated is now a major source of concern while there are no corresponding facilities to manage these wastes. The current waste disposal method which is landfilling is adequately insufficient and are at the end of its life span, with no intention of replacement due to space scarcity. Thus, this study aims to offer an overview of the issues connected with food waste generation and management in South Africa. Furthermore, this paper aims to summarize the result of acute food waste production on public health and the environment, and the economic and

---

* Corresponding Author's Email: olusolaolt@gmail.com; olusola@conserveafricainitiative.org.

In: Municipal Solid Waste Management and Improvement Strategies
Editor: Adam Fitz
ISBN: 979-8-88697-720-2

environmental gains of food waste management. Significant analysis of the present methods of food waste management and proposition for an improvement strategy were also discussed.

Keywords: food waste, low-income countries, landfilling, public health, South Africa

Introduction

Food waste (FW) can be defined as uneatable portions of food, any kind of food removed from the supply chain for the purpose of disposal or to be recovered (via composting, anaerobic digestion, or landfilling) (Aschemann-Witzelde HoogeAmaniBech-Larsen & Oostindjer, 2015). FW also includes waste foods that are processed into animal feeds (Aschemann-Witzelde HoogeAmaniBech-Larsen & Oostindjer, 2015). Food wastes (FWs), which are mostly organic, occur when more foods are bought or cooked than what is requires thus ending up in thrash/waste bins (British Columbia Ministry of Environment, 2015). Each time food wastage occurs, it connotes that all the resources (such as land, water, energy, manpower etc.) employed for its production have been wasted (Prescott et al., 2019; World Biogas Association, 2018). FW has become a topic of relevant discussion among policy makers and in the research community owing to its effect on public health and the environment (Filimonau et al., 2019; Kasza et al., 2019). Several studies have proposed the potential applications of FW for a sustainable economy and this include, gasification (Elkhalifa et al., 2019), compost conversion for soil amendment and bio-methane production as alternative sources of energy (Nayak & Bhushan, 2019; Yeo et al., 2019). A number of factors have been attributed as the causes of FW and these include: socio-demographic, behavioural and attitudinal factors etc. (Ramukhwatho et al., 2018). FW now seats at the peak position/summit of the municipal solid waste (MSW) streams based on most of the recent researches (AyeleruOkonta et al., 2016, 2018). Researches have shown that about one-third (~1.4 billion tonnes) of the total quantity of food manufactured globally becomes FW at various supply chains (Aschemann-Witzelde HoogeAmaniBech-Larsen & Gustavsson, 2015; Morone et al., 2019; Thyberg & Tonjes, 2016; Tsang et al., 2019).

In the high-income nations, FW occurs via the attitudes of consumers while in sub-Sahara Africa (SSA) or developing nations, it arises at the early phases of the food chain caused by inadequate storage facilities (Kasza et al.,

2019). For example, in South Africa the amount of food produced yearly is ~30 million tonnes, almost half of this quantity ends up as FW generated directly or indirectly. In utter variance, ~12 million people go to bed every night without food to eat. Thus, the economic value of FW generated in South Africa and its associated costs on the society when it is landfilled were estimated at ~USD$ 4 billion annually (Nahman et al., 2012). FW represents a huge financial burden on the nation of South Africa. This data is based on the study carried out by the Council for Scientific and Industrial Research (CSIR) of the Republic South Africa (RSA) which revealed that ~2% of the GDP of the country is lost in the form of FW in contrast with the previous report of 2012 that stated ~0.8% (Farmer's Weekly, 2017; Nahman et al., 2012; Nahman & De Lange, 2013; SEED, 2018). Based on these statistics, South Africa has been rated number one among the FW generators in Africa (Ramukhwatho et al., 2014; Reynolds, 2013). Established on the aforementioned stats, there is a need for urgent interventions to minimize FW impact vis-à-vis hunger and malnutrition on the general populace and to mitigate its effect on public health and the ecosystem (Dou et al., 2016). This present study centers on the significant public health challenges confronting members of the public based on the decisions they are taking unconsciously. This study also summarizes the mitigation measures and challenges of future work.

Global Food Wastes Production

People treat food as a disposable item both in the developed and developing countries and a research has shown that ~1.3 and ~2.1 billion tonnes of food meant for consumption becomes FW globally (Oelofse, 2014). FW currently occupies the largest portion of the municipal solid waste (MSW) but the quantity that are separately collected is ~2% of the FW fraction (FOE (Friends of the Earth), 2007). Globally, it has been estimated that the amount wasted yearly due to FW is ~USD$ 900 billion (Commonwealth of Australia, 2017a). A recent study has also shown that ~33% of the food manufactured universally turned to FW at various processing stages (Aschemann-Witzelde HoogeAmaniBech-Larsen & Gustavsson, 2015). The US environmental protection agency (EPA) has asserted that FW constitutes ~15% of the entire waste generated in the United States of America. Another research has revealed that ~5 million tonnes of FW are produced in Australia yearly and it has been valued at ~$USD 15 billion (Commonwealth of Australia, 2017a).

Recent research has shown that EU-28 produces ~100 million tonnes of FW yearly where households add little less than 50%. Some of the member states which include, Finland, Denmark, Norway and Sweden contribute ~20–30%

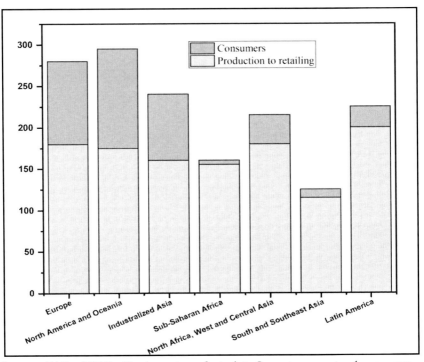

Source: FAO, (2013), Food wastage footprint: Impacts on natural resources - Summary report. (https://www.fao.org/3/i3347e/i3347e.pdf). Reproduced with permission.

Figure 1. Per capita food losses and waste (kg/year).

(Beretta et al., 2013; Xue et al., 2017). The bulk of these FWs ended up at the landfills thus negatively impacting public health, the environment and economic (McKenzie, 2015). FW which is a significant fraction (~30-40%) of the organic waste, when disposed of, to landfill sites (LSs), it generates methane which is an important source of greenhouse gases (GHGs) (Hall et al., 2009). The greenhouse gas (GHG) generated is ~21 times more stronger compared to carbon monoxide (CO) (New Mexico Recycling Coalition, 2014). The disposal of FW to landfills leads to heightening CO_2 emission to the atmosphere, thus adding to climate change and in the long run, giving rise to global warming (Hocke, 2014). The quantity of FW generated in Nigeria is

~40% of the total quantity of food produced. On a global scale, FW is having harmful effect on the environment, social and economy as its costs per year have been estimated to ~USD\$ 800 billion, USD\$ 910 billion and USD\$ 1 trillion respectively. Thus, the management of FW globally is a complex issue due to the number entities involved in the production, transportation, sales, and distribution and disposal (Commonwealth of Australia, 2017a). Figure 1 shows the distribution of FW on a global scale.

Description of Study Area

The Republic of South Africa (RSA) is a black dominated nation as well as multiethnic groups (Cook, 2020). South Africa (Figure 2) is situated in the sub-Saharan Africa (SSA) region, in southern tip on the African continent and stretches through -28.4792625 S and longitude 24.6727135 E (AyeleruOkonta et al., 2018; Ayeleru & Olubambi, 2021; Geodatos, 2021; McKenna, 2011). South Africa came into existence by the coming together of four British colonies in 1910, it became a Republic in 1961 and is usually referred to as the Republic of South Africa (RSA) (McKenna, 2011). RSA is seceded from nations like Namibia, Botswana, Zimbabwe, Mozambique, and Eswatini (Swaziland) from northwest to northeast and surrounds Lesotho with ~5000 km boundaries (Byrnes, 1996). The topography in RSA is both plain and flat highland and the coastal areas are thin, the climatical condition is usually dry, the daytime is usually sunny, and nighttime is frosty. Its land surface area is about ~1.2 million square kilometers (km^2). The major economic activities in RSA are mining, transport, energy, manufacturing, tourism, and agriculture. The current population of RSA is ~60 million with female occupying the highest percentage of ~51% while male occupies the rest (Government of South Africa, 2021). The continuous population growth in RSA has become a major contributing factor to the rapid FW generation (Ayeleru & Olubambi, 2021).

Food Waste Composition and Characterization Studies

An urgent intervention is required to achieve complete diversion of FW from landfills and these are through the physical and chemical composition studies of FW (EPA Sustainable Materials Management, 2014). FW physical composition study is important for reasons like: estimation of material recovery potential, discovery of source of waste constituents, acceleration of design of processing equipment; determination of physical, chemical and

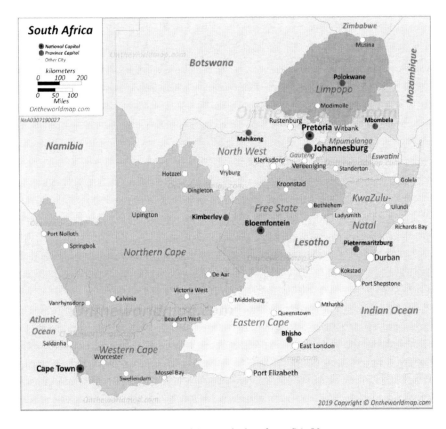

Figure 2. Map of South Africa. (With permission from SA-Venues.com; https://ontheworldmap.com/south-africa/).

thermal properties of waste and checking of conformity with both national and international standards (AyeleruNtuli et al., 2016a, 2016b, 2016c; Edjabou et al., 2015). The chemical composition data are also very crucial for the design of reactors and process stability of an anaerobic digestion. The data comprise of moisture content (MC), volatile solids content, nutrient content, particle size and biodegradability (Oliveira & Doelle, 2015). The anaerobic digestion of FW is one of the suitable techniques for recouping FW and biogas with volume of ~50-70% methane is produced from the process (Baky et al., 2014). Alternatively, composting fulfils the proposition for the complete diversion of

FW from landfills. It is environmentally desirable compared to landfilling. Composting is referred to as the aerobic decomposition of the organic fraction of municipal solid waste (OFMSW) in the presence of oxygen with the evolution of heat leading to temperature increase (Piñero, 2009; Sundberg, 2003). There are parameters which are very critical to composting process and they include, the carbon to nitrogen ratio (C/N), moisture content, pH, temperature, particle size, organic matter, ash content and total organic content (Román et al., 2015; Tiquia, 2005).

Table 1. Capacity details of landfill sites
in the City of Johannesburg (CoJ)

Dumping site	Remaining life of dump site (years)	Date of closure (month & year)
Marie Louise	—	1-Apr-21
Robinson Deep	—	1-Nov-21
Ennerdale	4.7	1-Sep-26
Goudkoppies	8	1-Feb-30

Figure 3. (a) Pictorial view of Robinson Deep landfill site; (b) Fruit waste been driven to the skip at the Joburg Fruit and Vegetable Market; (c) Fruit and vegetable waste at the skip waiting to be conveyed to Robinson Deep LS; (d), (e); Waste quantification activities at the Marie Louise LS and (f) Administering of questionnaire/waste education at Naledi extension in Soweto, South Africa.

Reliable data on FW quantification study in South Africa are very scanty and where they are available, they are usually unreliable due to the fact that this activity is not conducted regularly (Miezah et al., 2015). The main objective of a FW quantification (physical composition) study is to allow municipalities to evaluate the amount of FW generated over a period of one year in a country. Two of these researches were performed in the summer and winter of 2015 and 2016 in the City of Johannesburg (CoJ) to evaluate the amount of organic wastes as a proposition for the complete diversion of FW from landfill sites (LSs) and for FW to be channeled as a potential feed for biogas digester since all the four functional LSs in the CoJ (Table 1) would soon be closed (AyeleruOgundele et al., 2018; AyeleruOkonta et al., 2016, 2018; Zhang et al., 2007). From the Table 1, for instance, the Marie Louise and Robinson Deep LSs were expected to have been closed. Figure 3 shows waste quantification studies conducted alongside the waste pickers in 2015 and 2016 at the Marie Louise LS. It was also carried out at the fruit and vegetable market and questionnaires were administered to residents in one of the suburbs in Johannesburg on awareness creation vis-à-vis waste minimization, source separation and recycling.

Drivers of Food Wastes Generation

Drivers has been defined as the anthropogenic activities of people on the global environment resulting into negative impact on public health and the ecosystem (Contreras et al., 2010). While the causes of FW in the high-income countries have been attributed to affordability, accessibility, little or no knowledge on how food is cultivated, the causes in sub-Sahara Africa (SA) and particularly in South Africa is poor standard of living. A recent research has shown that unemployed citizens generate more FW since they mostly do all the cooking of their food at home and bulk of it is usually organic waste whereas the employed citizens consumed less organics but generate more inorganic wastes since they generally depend on eateries for their meals (AyeleruOkonta et al., 2016, 2018; Thyberg & Tonjes, 2016). From the research conducted by the Food and Agricultural Organization (FAO) on the distribution of FW on global scale (Figure 1), the amount of FW in Europe and America was ~280-300 kg/year and that from sub-Saharan Africa is 50% less compared to Europe and America (120-170 kg/year) (Al Seadi et al., 2013).

Climate Change and Human Health

~14 billion metric tonnes of carbon dioxide equivalents (CO_2eq) and ~30% anthropogenic greenhouse gases (GHGs) are generated from food supply chain (Poore & Nemecek, 2018). The bulk of this food turns to FWs which are biodegradable organic matters (Climate Council, 2016; Edell; Pearson). The disposal of FWs to landfills has a significant impact on human health and the environment since FW usually decomposes and releases methane into the atmosphere (Thyberg & Tonjes, 2016). Methane is an importance source of GHGs, usually about 21 times more stronger compared to carbon monoxide (CO) (New Mexico Recycling Coalition, 2014). When the concentrations of GHGs are distorted, countless health issues ensue in human and there is severe impact on the ecosystem (National Institute of Environmental Health Sciences). As the amount of CO_2 emission into the atmosphere increases, climate change and global warming occur (Hocke, 2014; Waqas et al., 2019). Climate change threatens the health of the public both locally and internationally in this 21st century (Doctors for the Environment Australia, 2016). The impacts of climate change endanger public health via its effect on food, water, air quality and weather (United States Environmental Protection Agency, 2017). Hence, tackling this issue of climate change regarding human health is a complicated problem since the decisions of man (which includes, cooking of excess food, wasting of the excess food, dumping of the wasted food at the landfills etc.) are difficult to curtail and these all affect the well-being of the public (Portier et al., 2010). The main drivers of climate change are the anthropogenic activities of man which give rise to substances like the tropospheric ozone (O_3), chlorofluorocarbons etc. (Bunyavanich et al., 2003; Gerardi & Kellerman, 2014; McMichael & Lindgren, 2011). Global warming is an occurrence that is well-known globally for its ability of posing threat to public health and the environment (American Lung Association of California, 2004). Figure 4 shows the impact of climate change on public health and on the five portions of the global environment (as indicated by the red- and purple-coloured lines). These aspects subsequently affect other environmental factors (such as, nutrient replenishment, air quality, species extinction, water quality etc.) (GREENTUMBLE, 2017). As shown in Figure 4, the consequences of climate change on public health are about twelve and they are mainly health impairments as indicated in the twelve rectangles with gold colours (Portier et al., 2010). Table 2 presents the sources of GHGs, and their consequences on public health and the environment.

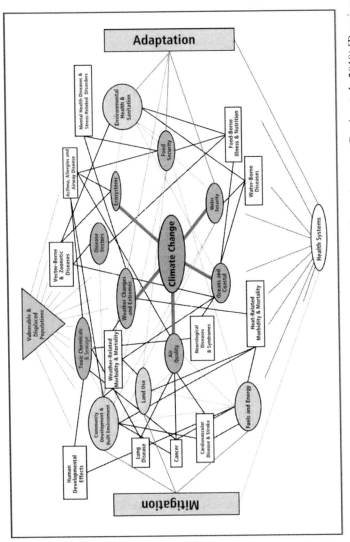

Figure 4. Straight impact of climate change on human health and five portions of the environment (Portier et al., 2010) [Requires no permission as it lies in public domain]. Five characteristics of the human environment (red thick lines, purple circles (Toxic Chemicals & Sewage, Weather Changes and Extremes, Disease Vectors, Ecosystems, Food Security, Water Security, Oceans and Coastal, Air Quality) that are directly impacted by climate change also have an impact on other environmental parameters. Twelve different facets of human health are then affected by these environmental changes (small tan boxes). To combat climate change, mitigation and adaptation modify the human environment, which thus modifies human health.

Table 2. Sources of greenhouse gases, health consequence
on human health and impact on the environment

Greenhouse gases	Human health consequences of climate change	Impact of climate change on the environment
Carbon dioxide	Asthma	Sea-level rise
Methane	Respiratory allergies	Flooding and drought
Nitrous oxide	Airway diseases	Heat waves
Tropospheric ozone	Cancer	Hurricanes and storms
Chlorofluorocarbons	Cardiovascular disease and stroke	Poor air quality
Carbon monoxide	Foodborne diseases and nutrition	Changes in water availability
Water vapor	Heat-related morbidity and mortality	Air pollution
Hydrofluorocarbons	Human developmental effects	Distorted rainfall patterns
Perfluorocarbons	Mental health and stress-related disorders	Reduction in precipitation
Sulfur hexafluoride	Neurological diseases and disorders	
	Vector-borne and Zoonotic Diseases	
	Waterborne Diseases	
	Weather-Related Morbidity and Mortality	

(American Lung Association of California, 2004; Commonwealth of Australia, 2006, 2017b; Enríquez-de-Salamanca et al., 2017; Gerardi & Kellerman, 2014; McMichael et al., 2003; OECD, 2008; Portier et al., 2010; Repetto, 2012; Sovacool, 2014; Watkiss et al., 2005; Wolf & Menne, 2007)

Economic Benefits of Food Waste Management

FW has become an issue of main concern in all economies globally (Venkat, 2011). According to UN report, it is expected that the quantity of FW generated will drop by ~50% by the year 2030 (Marchant, 2021; Rashid & Shahzad, 2021), translating to a reduction of ~1.5 billion tonnes of CO_2 emissions (Robertson-Fall, 2021). The amount of FW generated yearly is ~930 million tonnes with ~60% originating for residential areas, ~25% from food waste services and ~15% emanating from retail services (Marchant, 2021). Another study has estimated the amount of FW generated globally to be ~33% of the total food produced globally, and its associated cost is valued at USD$ 1 trillion per year (Tyagi, 2020). It therefore becomes very imperative to put a system in place for the reduction of FW. When FW is reduced, it translates to cost saving. So, for each dollar paid on FW reduction, ~USD$ 15 is saved by industries (Harned, 2017). Also, when FW is reduced, a significant saving is achieved because bills of food for households are lowered and costs of disposal by eateries, farmers and food processing industries are equally slashed (Bell, 2012; Parry et al., 2015). When FW is properly managed, it leads to creation of jobs in composting facilities and biomethane generation plants. A study has shown that about 15% additional jobs are created from processes like composting etc. (Goldstein, 2014). Composting of food waste is one of the safest methods of managing wet waste. It is also known as aerobic process in which complex organic materials break down via the activity of microorganisms to produce composts for soil amendment or conditioning. Composts are packaged and sold to farmers for the growth of the crops, and this also serves as a phase to transiting from linear model of material utilization to circular economy (CE). Through composting process, life of landfill facilities are extended and costs of construction of new disposal facilities are saved (Ayilara et al., 2020; Rashid & Shahzad, 2021).

Environmental Benefits of Food Waste Management

FW contributes to environmental issues, which includes, global warming, impairment to ecosystem and diminution of natural resources (Ayeleru et al., 2020; Diaz, 2011). When more food is cooked or bought beyond what is required, it eventually ends up in the bins and turns to FW. This waste food, when dump at the disposal facilities as is mostly common and as the major method of FW disposal in South Africa, it degrades and produces methane which goes into the atmosphere and is a powerful source of greenhouse gases (GHGs). FW has been known to contribute to ~8% GHGs, a major source of global warming (Nicastro & Carillo, 2021; Seberini, 2020). When FW is properly managed rather than being disposed of into landfills, methane emissions from landfill facilities will be minimized, carbon footprint will be reduced, energy and resources will be conserved, and pollution will be prevented (EPA United States Environmental Protection Agency, 2021; Food and Agriculture Organization of the United Nations, 2019). Carbon footprint is the sum of the quantity of GHGs that are emitted by a material throughout its life cycle and is usually expressed in kilograms of CO_2 equivalents (FAO, 2013).

Mitigation Measures to Reduce the Negative Impact of Food Waste on Public Health and the Environment

To achieve reduction of the impact of climate change on human health and the environment, the first and most prominent means is reprocessing or eating of leftover food since it has been proven to be one of the most successful means to reduce FW at households (Schanes et al., 2018). Distribution to the needy instead of allowing food to go bad is also a better way (Parry et al., 2015). Education or awareness campaign can follow (American Public Health Association, 2001; Clarke, 2014; Lamb & Fountain, 2010). Since members of the public are concerned about their health, especially the health impairment caused by landfilling of FW, this should be the main focus of the awareness campaign (Walter, 2014). As part of the education campaign, municipalities can put a team in place who will be educating households within the suburbs on the need for FW reduction, the need to start home composting and the need to increase recycling rates (AyeleruOkonta et al., 2016). Incorporating FW reduction strategy into basic and higher education curriculum can also serve as guideline for policy maker. Separation of FW at source should as well be encouraged. This will involve the use of different branded plastic bags or waste bins of varied colours for storage of different waste types before collection (Mason et al., 2011). Government can invest in biogas and composting plants so that FW from source separation project can serve as feeds for biogas digester and composting plant (Waste Authority, 2014). Grocery stores, fruit and vegetable market and charitable establishments can enter partnership on alleviating FW. Rather than allowing wastage, FW or fruit and vegetable waste (FVW) can be donated to charity organizations or non-governmental organizations (NGOs) who will in turn donate them to public or local authorities. Also, incentive model can be created for citizens who are actively participating in source separation, home composting and various methods of FW management other than landfilling which is the prominent method of managing FW in South Africa. Data can be normalized to evaluate FW and track FW reduction yearly. Finally, government can lend support to farmers by assisting them in the building of modern storage facilities to avoid food wastage.

Challenges and Future Research

Reducing the impact caused by climate change on health of the public and the global environment is an issue of serious concern. Although, the City of Johannesburg has begun a project on the generation of biomethane from organic waste as a means for complete diversion of FW from landfills, but the project is still at an early development stage (Niklasson & Bergquist Skogfors, 2018). The challenges of FW management are so enormous that it is beyond what the government alone can deal with it. Hence, to win a total war over the issue of climate change caused by poor FW disposal resulting into different health issues and environmental problems, the co-operation of all stakeholders (including, members of the public, local authorities, policy makers, researchers, waste management agencies etc.) is very crucial (AyeleruOkonta et al., 2018). When all the stakeholders agree, the challenges of food uncertainty, economic wastefulness; and environmental deterioration which have been cited as the major contributing factors to health issue in human will be addressed. The public will also be eager to participate in source separation of waste and they would begin to have a change of attitude towards climate change (Warshawsky, 2019). Currently, source separation of waste is not very functional in all the provinces in South Africa, however, efforts can be made so that the public can embrace this program to minimize health challenges confronting members of the public through their actions and inactions. It is high time we know that our continuous existence on earth and the sustainability of our ecosystem depend solely on our attitudes towards climate change.

Conclusion

Increasing population and economic growth has been ascribed as the causes of severe FW generated both nationally and internationally. The bulk of FW is deposited at the landfills in South Africa and when it degrades, harmful substances are released into the atmosphere which present risk to health of the public and the ecological environment. With adequate funding for waste management and effective source separation program in place, this waste can be completely diverted from landfills and resource such as composts for soil amendments and biogas for cooking and heating will be recouped from the waste and will also serve as a source of revenue generation. The unemployed youths will be gainfully employed when composting plants and biodigesters for biomethane generation are in operation in South Africa.

References

Al Seadi, T., Owen, N., Hellström, H., & Kang, H. (2013). Source separation of MSW: an overview of the source separation and separate collection of the digestible fraction of household waste, and of other similar wastes from municipalities, aimed to be used as feedstock for anaerobic digestion in biogas plants. *IEA Bioenergy*, 7-10.

American Lung Association of California. (2004). *FACT SHEET: Air Quality and Health Impacts of Greenhouse Gas Emissions and Global Warming*. Retrieved from http://www.dnrec.delaware.gov/dwhs/Info/Regs/Documents/alac_impacts_fs.pdf

American Public Health Association. (2001). *Climate change: mastering the public health role, a practical guidebook*. In: APHA.

Aschemann-Witzel, J., de Hooge, I., Amani, P., Bech-Larsen, T., & Gustavsson, J. (2015). Consumers and food waste-a review of research approaches and findings on point of purchase and in-household consumer behaviour. *Agricultual and Applied Economics Association*.

Aschemann-Witzel, J., de Hooge, I., Amani, P., Bech-Larsen, T., & Oostindjer, M. (2015). Consumer-related food waste: causes and potential for action. *Sustainability, 7*(6), 6457-6477.

Ayeleru, O. O., Dlova, S., Akinribide, O. J., Ntuli, F., Kupolati, W. K., Marina, P. F., Blencowe, A. & Olubambi, P. A. (2020). Challenges of plastic waste generation and management in sub-Saharan Africa: A review. *Waste management, 110*, 24-42.

Ayeleru, O. O., Ntuli, F., & Mbohwa, C. (2016a). *Characterisation of Fruits and Vegetables Wastes in the City of Johannesburg*. Paper presented at the Proceedings o (the World Congress on Engineering and Computer Science.

Ayeleru, O. O., Ntuli, F., & Mbohwa, C. (2016b). *Municipal Solid Waste Composition Determination in the City of Johannesburg*. Paper presented at the Proceedings of the World Congress on Engineering and Computer Science.

Ayeleru, O. O., Ntuli, F., & Mbohwa, C. (2016c). *Utilization of Organic Fraction of Municipal Solid Waste (OFMSW) as Compost: A Case Study of Florida, South Africa*. Paper presented at the Proceedings of the World Congress on Engineering and Computer Science.

Ayeleru, O. O., Ogundele, O. M., Gbashi, S., Adeniran, J. A., Dlova, S., Kayitesi, E., Ntuli, F., N., Belaid, M. Mbohwa, C. & Olubambi, P. A. (2018). Fruit and vegetable consumption: A case study of food culture vis-à-vis health awareness among the students of the University of Johannesburg, South Africa. In D. N. Motohashi (Ed.), *Fruit and Vegetable Consumption and Health: New Research*. New York, USA: Nova Science Publishers.

Ayeleru, O. O., Okonta, F. N., & Ntuli, F. (2016). *Characterization, management and utilization of landfill municipal solid waste: a case study of Soweto*. (Masters Dissertation). University of Johannesburg, Johannesburg.

Ayeleru, O. O., Okonta, F. N., & Ntuli, F. (2018). Municipal solid waste generation and characterization in the City of Johannesburg: A pathway for the implementation of zero waste. *Waste management, 79*, 87-97.

Ayeleru, O. O., & Olubambi, P. A. (2021). Solid Waste Treatment Processes and Remedial Solution in the Developing Countries. In R. R. K. Dr, P. G. Ravindran, & P. M. H. Dehghani (Eds.), *Soft Computing Techniques in Solid Waste and Wastewater Management* (pp. 233-246). United States: Elsevier.

Ayilara, M. S., Olanrewaju, O. S., Babalola, O. O., & Odeyemi, O. (2020). Waste management through composting: Challenges and potentials. *Sustainability, 12*(11), 4456.

Baky, M. A. H., Khan, M. N. H., Kader, M. F., & Chowdhury, H. A. (2014). Production of Biogas by Anaerobic Digestion of Food waste and Process Simulation. Paper presented at the *ASME 2014 8th International Conference on Energy Sustainability collocated with the ASME 2014 12th International Conference on Fuel Cell Science, Engineering and Technology*.

Bell, R. (2012, 27 March 2012). *Reducing food waste has economic, environmental and social benefits*. Retrieved from https://www.canr.msu.edu/news/reducing_food_waste_has_economic_environmental_and_social_benefits.

Beretta, C., Stoessel, F., Baier, U., & Hellweg, S. (2013). Quantifying food losses and the potential for reduction in Switzerland. *Waste management, 33*(3), 764-773.

British Columbia Ministry of Environment. (2015). *Residential Food Waste Prevention*. Retrieved from Vancouver: https://www2.gov.bc.ca/assets/gov/environment/waste-management/recycling/organics/resources/food_waste_reduction_toolkit.pdf

Bunyavanich, S., Landrigan, C. P., McMichael, A. J., & Epstein, P. R. (2003). The impact of climate change on child health. *Ambulatory pediatrics, 3*(1), 44-52.

Byrnes, R. M. (1996). *South Africa: A country study*. Washington: U.S Library of Congress.

Clarke, C. (2014). Reducing Food Waste: Recommendations to the 2015 Dietary Guidelines Advisory Committee. *Diss. Boston, Massachusetts: Simmons College.* Retrieved from https://health.gov/dietaryguidelines/dga2015/comments/uploads/CID430_Tufts_University-_Reducing_Food_Waste-_DGAC_Comment.pdf.

Climate Council. (2016, 16 October 2016). *From Farm to Plate to the Atmosphere: Food-Related Emissions*. Retrieved from https://www.climatecouncil.org.au/from-farm-to-plate-to-the-atmosphere-reducing-your-food-related-emissions/

Commonwealth of Australia. (2006). *Climate change impacts & risk management: a guide for business and government*: Australian Greenhouse Office, Department of the Environment and Heritage.

Commonwealth of Australia. (2017a). *National Food Waste Strategy Halving Australia's Food Waste By 2030*. Retrieved from https://www.environment.gov.au/system/files/resources/4683826b-5d9f-4e65-9344-a900060915b1/files/national-food-waste-strategy.pdf.

Commonwealth of Australia. (2017b). *Working together to reduce food waste in Australia*. Retrieved from Australia https://www.environment.gov.au/system/files/resources/29c0f94d-92ac-44d7-ac43-3051cff75162/files/food-waste-fact-sheet.pdf.

Contreras, F., Ishii, S., Aramaki, T., Hanaki, K., & Connors, S. (2010). Drivers in current and future municipal solid waste management systems: cases in Yokohama and Boston. *Waste Management & Research, 28*(1), 76-93.

Cook, N. (2020). *South Africa: Current issues, economy, and US relations*. In: Congressional Research Service.

Diaz, L. F. (2011). The 3Rs as the Basis for Sustainable Waste Management: Moving Towards Zero Waste. *Paper presented at the Third Regional 3Rs Forum in Asia and the Pacific*: Singapore.

Doctors for the Environment Australia. (2016). *Climate Change & Health in Australia: Fact Sheet*. Retrieved from https://www.dea.org.au/wp-content/uploads/2017/02/DEA_Climate_Change__Health_Fact_Sheet_final.pdf

Dou, Z., Ferguson, J. D., Galligan, D. T., Kelly, A. M., Finn, S. M., & Giegengack, R. (2016). Assessing US food wastage and opportunities for reduction. *Global Food Security, 8*, 19-26.

Edell, B. (26 JULY 2016). *Food Waste Contributes to Greenhouse Gas Emissions More Than You Would Think!* Retrieved from https://www.globalgreen.org/blog/food wasteequalsgreenhousegas

Edjabou, M. E., Jensen, M. B., Götze, R., Pivnenko, K., Petersen, C., Scheutz, C., & Astrup, T. F. (2015). Municipal solid waste composition: Sampling methodology, statistical analyses, and case study evaluation. *Waste Management, 36*, 12-23.

Elkhalifa, S., Al-Ansari, T., Mackey, H. R., & McKay, G. (2019). Food waste to biochars through pyrolysis: A review. *Resources, Conservation and Recycling, 144*, 310-320.

Enríquez-de-Salamanca, Á., Díaz-Sierra, R., Martín-Aranda, R. M., & Santos, M. J. (2017). Environmental impacts of climate change adaptation. *Environmental Impact Assessment Review, 64*, 87-96.

EPA Sustainable Materials Management. (2014). *A Guide to Conducting and Analyzing a Food Waste Assessment*. Retrieved from https://www.epa.gov/sites/production/files/2015-08/documents/r5_fd_wste_guidebk_020615.pdf

EPA United States Environmental Protection Agency. (2021, 2 July 2021). *Reducing Wasted Food At Home*. Retrieved from https://www.epa.gov/recycle/reducing-wasted-food-home#:~:text=Benefits%20of%20Reducing%20Wasted%20Food,-Helpful%20Links&text=Reduces%20methane%20emissions%20from%20landfills, waste%20and%20then%20landfilling%20it).

FAO. (2013). *Food wastage footprint: Impacts on natural resources - Summary report*. Retrieved from https://www.fao.org/3/i3347e/i3347e.pdf

Farmer's Weekly. (2017, 2018). *The cost of South Africa's food loss and waste*. Retrieved from https://www.farmersweekly.co.za/opinion/by-invitation/cost-south-africas-food-loss-waste/

Filimonau, V., Krivcova, M., & Pettit, F. (2019). An exploratory study of managerial approaches to food waste mitigation in coffee shops. *International Journal of Hospitality Management, 76*, 48-57.

FOE (Friends of the Earth). (2007). *Food waste collections* Retrieved from London: https://friendsoftheearth.uk/sites/default/files/downloads/food_waste.pdf

Food and Agriculture Organization of the United Nations. (2019). *The state of food and agriculture: Moving forward on food loss and waste reduction.* Retrieved from https://www.fao.org/3/ca6122en/ca6122en.pdf.

Geodatos. (2021). *South Africa Geographic coordinates.* Retrieved from https://www.geodatos.net/en/coordinates/south-africa.

Gerardi, D. A., & Kellerman, R. A. (2014). Climate change and respiratory health. *Journal of occupational and environmental medicine, 56,* S49-S54.

Goldstein, J. (2014). *From Waste to Jobs: What Achieving 75 Percent Recycling Means for California.* Retrieved from https://www.nrdc.org/sites/default/files/green-jobs-ca-recycling-report.pdf.

Government of South Africa. (2021). *South Africa at a glance.* Retrieved from https://www.gov.za/about-sa/south-africa-glance.

Greentumble. (2017, 21 May 2017). *Grave Effects of Climate Change on the Environment.* Retrieved from https://greentumble.com/grave-effects-of-climate-change-on-the-environment/.

Hall, K. D., Guo, J., Dore, M., & Chow, C. C. (2009). The progressive increase of food waste in America and its environmental impact. *PloS one, 4*(11), e7940.

Harned, C. (2017, 8 March 2017). *The Economic Benefits of Reducing Food Waste.* Retrieved from https://cuer.law.cuny.edu/?p=1910

Hocke, L. (2014). *Encourage Food Waste Reduction.*

Kasza, G., Szabó-Bódi, B., Lakner, Z., & Izsó, T. (2019). Balancing the desire to decrease food waste with requirements of food safety. *Trends in Food Science & Technology, 84,* 74-76.

Lamb, G., & Fountain, L. (2010). *An investigation into food waste management.* Retrieved from http://www.actiondechets.fr/upload/medias/group_b_report_compressed.pdf.

Marchant, N. (2021, 26 March 2021). *The world's food waste problem is bigger than we thought - here's what we can do about it.* Retrieved from https://www.weforum.org/agenda/2021/03/global-food-waste-solutions/.

Mason, L., Boyle, T., Fyfe, J., Smith, T., & Cordell, D. (2011). *National food waste data assessment: final report.* Retrieved from Australia:

McKenna, A. (2011). *The history of southern Africa.* New York: Britannica Educational Publishing.

McKenzie, M. (2015). *South Africa generates over 9 million tonnes of food waste annually.* Retrieved from http://www.urbanearth.co.za/articles/south-africa-generates-over-9-million-tonnes-food-waste-annually.

McMichael, A. J., Campbell-Lendrum, D. H., Corvalán, C. F., Ebi, K. L., Githeko, A., Scheraga, J. D., & Woodward, A. (2003). *Climate change and human health: risks and responses*: World Health Organization.

McMichael, A. J., & Lindgren, E. (2011). Climate change: present and future risks to health, and necessary responses. *Journal of internal medicine, 270*(5), 401-413.

Miezah, K., Obiri-Danso, K., Kádár, Z., Fei-Baffoe, B., & Mensah, M. Y. (2015). Municipal solid waste characterization and quantification as a measure towards effective waste management in Ghana. *Waste management, 46,* 15-27.

Morone, P., Koutinas, A., Gathergood, N., Arshadi, M., & Matharu, A. (2019). Food waste: Challenges and opportunities for enhancing the emerging bio-economy. *Journal of Cleaner Production.*

Nahman, A., De Lange, W., Oelofse, S., & Godfrey, L. (2012). The costs of household food waste in South Africa. *Waste Management, 32*(11), 2147-2153.

Nahman, A., & De Lange, W. J. (2013). *Costs of food waste along the value chain: Evidence from South Africa*: Presentation.

National Institute of Environmental Health Sciences. (25 September 2018). *Health Impacts.* Retrieved from https://www.niehs.nih.gov/research/programs/geh/climatechange/health_impacts/index.cfm.

Nayak, A., & Bhushan, B. (2019). An overview of the recent trends on the waste valorization techniques for food wastes. *Journal of environmental management, 233,* 352-370.

New Mexico Recycling Coalition. (2014). *Managing Food Waste in NM Restaurants.* Retrieved from http://www.recyclenewmexico.com/pdf/food-waste-management-restaurantsweb.pdf.

Nicastro, R., & Carillo, P. (2021). Food loss and waste prevention strategies from farm to fork. *Sustainability, 13*(10), 5443.

Niklasson, J., & Bergquist Skogfors, L. (2018). *Can organic waste fuel the buses in Johannesburg?: A study of potential, feasibility, costs and environmental performance of a biomethane solution for public transport.* (Master's thesis). Linköping University, Sweden. (LIU-IEI-TEK-A--18/03036—SE).

OECD. (2008). *Climate Change Mitigation« WHAT DO WE DO?* Retrieved from https://www.oecd.org/environment/cc/41751042.pdf.

Oelofse, S. H. (2014). *Food waste in South Africa: Understanding the magnitude, water footprint and cost.* In: Alive2green.

Oliveira, F., & Doelle, K. (2015). Anaerobic digestion of food waste to produce biogas: a comparison of bioreactors to increase methane content–a Review. *J Food Process Technol, 6*(8), 1-3.

Parry, A., James, K., & LeRoux, S. (2015). *Strategies to achieve economic and environmental gains by reducing food waste.* Retrieved from Banbury: https://newclimateeconomy.report/workingpapers/wp-content/uploads/sites/5/2016/04/WRAP-NCE_Economic-environmental-gains-food-waste.pdf.

Pearson, P. (26 September 2018). *Reducing food waste could dramatically cut GHG emissions.* Retrieved from https://www.greenbiz.com/article/reducing-food-waste-could-dramatically-cut-ghg-emissions.

Piñero, J. C. (2009). *Composting Food Waste as an Alternative to Landfill Disposal.* Month.

Poore, J., & Nemecek, T. (2018). Reducing food's environmental impacts through producers and consumers. *Science, 360*(6392), 987-992.

Portier, C. J., Thigpen, T. K., Carter, S. R., Dilworth, C. H., Grambsch, A. E., Gohlke, J., Hess, J. J., Howard, S., Luber, G., Lutz, J., Maslak, T., Radtke, M., Rosenthal J. P., Rowles, T., Sandifer, P., Scheraga, J., Schramm, P. J., Strickman, D., Trtanj, J. M., & Whung, P. Y. (2010). *A Human Health Perspective On Climate Change: A Report Outlining the Research Needs on the Human Health Effects of Climate Change.*

Retrieved from Research Triangle Park, NC: file:///C:/Users/user/Downloads/A%20
Human%20Health%20Perspective%20On%20Climate%20Change%20Full%20Rep
ort.pdf.

Prescott, M. P., Herritt, C., Bunning, M., & Cunningham-Sabo, L. (2019). Resources,
Barriers, and Tradeoffs: A Mixed Methods Analysis of School Pre-Consumer Food
Waste. *Journal of the Academy of Nutrition and Dietetics*.

Ramukhwatho, F., duPlessis, R., & Oelofse, S. (2018). Preliminary drivers associated with
household food waste generation in South Africa. *Applied Environmental Education
& Communication, 17*(3), 254-265.

Ramukhwatho, F. R., Du Plessis, R., & Oelofse, S. H. H. (2014). *Household food wastage
in a developing country: A case study of Mamelodi Township in South Africa*.

Rashid, M. I., & Shahzad, K. (2021). Food waste recycling for compost production and its
economic and environmental assessment as circular economy indicators of solid waste
management. *Journal of Cleaner Production, 317*, 128467.

Repetto, R. (2012). Economic And Environmental Impacts of Climate Change In Virginia.
Demos, New York, 11.

Reynolds, C. J. (2013). *Quantification of Australian Food Wastage with Input Output
Analysis*. Doctor of philosophy (applied mathematics), University of South Australia.

Robertson-Fall, T. (2021, 24 February 2021). *Five benefits of a circular economy for food*.
Retrieved   from   https://ellenmacarthurfoundation.org/articles/five-benefits-of-a-
circular-economy-for-food

Román, P., Martínez, M. M., & Pantoja, A. (2015). *Farmer´s Compost Handbook:
Experiences in Latin America*. Retrieved from Santiago: http://www.fao.org/3/a-
i3388e.pdf

Schanes, K., Dobernig, K., & Gözet, B. (2018). Food waste matters-A systematic review
of household food waste practices and their policy implications. *Journal of Cleaner
Production, 182*, 978-991.

Seberini, A. (2020). Economic, social and environmental world impacts of food waste on
society and Zero waste as a global approach to their elimination. *Paper presented at
the SHS Web of Conferences*.

SEED. (2018). *Waste to Food: Creating economic opportunities by recycling food waste*.
Retrieved   from   South   Africa: https://www.seed.uno/images/casestudies/SEED_
Case_Study_Waste_to_Food_South_Africa.pdf

Sovacool, B. K. (2014). Environmental issues, climate changes, and energy security in
developing Asia. *Asian Development Bank Economics Working Paper Series*(399),
17-14.

Sundberg, C. (2003). *Food waste composting: effects of heat, acids and size*: Sveriges
lantbruksuniv.

Thyberg, K. L., & Tonjes, D. J. (2016). Drivers of food waste and their implications for
sustainable policy development. *Resources, Conservation and Recycling, 106*, 110-
123.

Tiquia, S. M. (2005). Microbiological parameters as indicators of compost maturity.
Journal of applied microbiology, 99(4), 816-828.

Tsang, Y. F., Kumar, V., Samadar, P., Yang, Y., Lee, J., Ok, Y. S., Song, H., Kim, K-H., Kwon, E. E., & Jeon, Y. J. (2019). Production of bioplastic through food waste valorization. *Environment international, 127*, 625-644.

Tyagi, H. (2020, 16 October 2020). *World Food Day 2020: How food waste affects the economy.* Retrieved from https://www.timesnownews.com/business-economy/industry/article/world-food-day-2020-how-food-waste-affects-the-economy/667896

United States Environmental Protection Agency. (2017, 13 January 2017). *Climate Impacts on Human Health.* Retrieved from https://19january2017snapshot.epa.gov/climate-impacts/climate-impacts-human-health_.html

Venkat, K. (2011). The climate change and economic impacts of food waste in the United States. *International Journal on Food System Dynamics, 2*(4), 431-446.

Walter, T. (2014). *Impacts of Climate Change on Public Health in Australia: Recommendations for New Policies and Practices for Adaptation Within the Public Health Sector*: Deeble Institute.

Waqas, M., Nizami, A. S., Aburiazaiza, A. S., Barakat, M. A., Asam, Z. Z., Khattak, B., & Rashid, M. I. (2019). Untapped potential of zeolites in optimization of food waste composting. *Journal of environmental management, 241*, 99-112.

Warshawsky, D. N. (2019). The Challenge of Food Waste Governance in Cities: Case Study of Consumer Perspectives in Los Angeles. *Sustainability, 11*(3), 847.

Waste Authority. (2014). *Source Separation of Waste: Position Statement*. Retrieved from https://www.wasteauthority.wa.gov.au/media/files/documents/Source_Separation_of _Waste_2014.pdf

Watkiss, p., Downing, t., Handley, C., & Butterfield, R. (2005). *The Impacts and Costs of Climate Change.* Retrieved from https://ec.europa.eu/clima/sites/clima/files/strategies/2020/docs/final_report2_en.pdf

Wolf, T., & Menne, B. (2007). *Environment and health risks from climate change and variability in Italy.* Retrieved from Denmark:

World Biogas Association. (2018). *Global Food Waste Management: An Implementation Guide for Cities.* Retrieved from London: http://www.worldbiogasassociation.org/wp-content/uploads/2018/05/Global-Food-Waste-Management-Full-report-pdf.pdf

Xue, L., Liu, G., Parfitt, J., Liu, X., Van Herpen, E., Stenmarck, Å., O'Connor, C., Östergren, K., & Cheng, S. (2017). Missing food, missing data? A critical review of global food losses and food waste data. *Environmental science & technology, 51*(12), 6618-6633.

Yeo, J., Oh, J., Cheung, H. H. L., Lee, P. K. H., & An, A. K. (2019). Smart Food Waste Recycling Bin (S-FRB) to turn food waste into green energy resources. *Journal of environmental management, 234*, 290-296.

Zhang, R., El-Mashad, H. M., Hartman, K., Wang, F., Liu, G., Choate, C., & Gamble, P. (2007). Characterization of food waste as feedstock for anaerobic digestion. *Bioresource technology, 98*(4), 929-935.

Index